The Economic Geography of the UK

SAGE has been part of the global academic community since 1965, supporting high quality research and learning that transforms society and our understanding of individuals, groups, and cultures. SAGE is the independent, innovative, natural home for authors, editors and societies who share our commitment and passion for the social sciences.

Find out more at: **www.sagepublications.com**

The **Economic Geography**
of the **UK**

Edited by Neil M. Coe & Andrew Jones

Los Angeles | London | New Delhi
Singapore | Washington DC

First published 2010

SAGE Publications Ltd
1 Oliver's Yard
55 City Road
London EC1Y 1SP

SAGE Publications Inc.
2455 Teller Road
Thousand Oaks, California 91320

SAGE Publications India Pvt Ltd
B 1/I 1 Mohan Cooperative Industrial Area
Mathura Road, Post Bag 7
New Delhi 110 044

SAGE Publications Asia-Pacific Pte Ltd
33 Pekin Street #02-01
Far East Square
Singapore 048763

Library of Congress Control Number 2010930940

British Library Cataloguing in Publication data
A catalogue record for this book is available from the British Library

ISBN 978-1-84920-089-9
ISBN 978-1-84920-090-5 (pbk)

Typeset by C&M Digitals (P) Ltd, Chennai, India
Printed by CPI Antony Rowe, Chippenham, Wiltshire
Printed on paper from sustainable resources

CONTENTS

ACKNOWLEDGEMENTS

This book started out as an all-day series of sessions on the *New Geographies of the UK Economy* at the Royal Geographical Society with Institute of British Geographers (RGS-IBG) Annual Conference in London in August 2008. The sessions were co-organised with Mia Gray (University of Cambridge) who we thank for her enthusiasm in helping to get this project off the ground. We also thank all those who presented in, or attended, the conference sessions and convinced us that this project was viable! The chapters in the book build upon, but do not correspond exactly with, the conference session line-up and topics. We would like to thank all the contributors for their enthusiasm and commitment to this book, and for producing what we think are uniformly stimulating, timely and important commentaries on different aspects of the UK space-economy. Many thanks are also due to University of Manchester cartographer Graham Bowden for producing all the figures with his customary style and efficiency.

All proceeds from this book will go to the RGS-IBG's Economic Geography Research Group (EGRG). As such, they will hopefully go someway to replacing the revenue stream from Roger Lee and Jane Will's (1997) edited collection, *Geographies of Economies* (London: Arnold) which for many years provided very welcome funds to the research group. We would like to thank all the chapter authors for agreeing to this arrangement. For more on the ongoing activities of the EGRG, please visit: www.egrg.org.uk.

Finally, we would like to thank Robert Rojek and Sarah-Jayne Boyd at Sage for their unwavering support for this project, and Katherine Haw for her excellent production editing.

Neil M. Coe, Glossop (EGRG Chair, 2006–09)
Andrew Jones, Fulham (EGRG Chair, 2009–2012)
December 2009

PREFACE

This is an opportune moment for this collection of essays to appear. Thirty years ago, in the decade of the 1980s, what we have come to call, in sometimes inadequate shorthand, 'neoliberalism' muscled its way to dominance over the economy, the economic geography, and the polity more generally, of the United Kingdom. Over the period in which this book has been written, that dominance has been shaken, even challenged. Each distinct era in a society's history – 'distinct' in terms of the political and social understandings that underpin it – produces a particular geography. This is a moment, then, in which it is possible to assess the geographies produced in this most recent era in the UK, and to reflect upon the problems that lie within it and the challenges that its potential fracture throw up.

Moreover, these geographies are not only the *product* of particular socioeconomic formations, they are an active part of their very dynamics. This book documents in a host of ways how the acute inequalities generated over the last 30 years are reinforced in their effect by their geographical form (the chasm of the North–South divide in England being one of the most evident of these). And this spatial patterning of inequality can further divide a supposedly united kingdom. Or again, there is the argument that the dominance of an economy by a global financial centre can have deleterious macro economic effects, from pressures on costs to the importation of volatility. Or yet again, as we see in another chapter, geographical divides can disrupt the functioning of labour markets thereby reducing precisely the 'market efficiency' of which the proponents of this neoliberal period have made so much. In other words, the economic geography of a society is integral to the way that society functions, to its reproduction, and to its potential fracturing. We can see that clearly in a number of the contributions to this book.

The editors are therefore right to call in their Introduction for the reinvigoration of an economic geography that investigates these fundamental relations. Beyond the studies of the micro practices of firms and their managements, their changing cultures and their location decisions, we need also an understanding of how the shape of an economy and society is utterly bound up with its spatial organisation. The articles collected here begin to give us materials for that kind of investigation.

Such an investigation requires also that issues of class and of power are recognised as fundamental, not just to the way in which our economy works but also to our national geography. Economic geography in this guise must range widely, through issues of labour markets, housing, gender, ethnicity, health – even to the length of time we are likely to live. In this guise, too, it recognises that few of the processes it engages with are simply 'economic'. Whether it be government decisions over the location of major research facilities, or the location of civil servants, whether it be more general political commitments about the shape of the economy (accepting the dominance of market forces, encouraging the dominance of the financial sector and its associated services), whether it be the pushes (both political and cultural) towards changes in the tenurial structure within housing, economic geography is utterly bound up with the more general governance of the country.

The editors point out that the book can be approached in a number of different ways (you don't have to read it from first chapter to last, in that order, to learn from it). I agree. One way to work with the book might be to follow through it themes which crop up persistently in the different chapters. The most evident of these is the theme of inequality, its persistence and sharpening, the host of dimensions through which it is produced and experienced, and its intersecting – and often overlapping and mutually reinforcing – geographies. What kind of a society are we building here? Another theme might be the financial sector, with its iconic geographical centerpiece in the square mile of the City of London. The growth of this sector, and also the dissemination of its mores, has been a central and crucial feature of the era on which this book reflects. A number of chapters address this issue from a variety of angles – debates over how to understand 'the power' of the City and finance, and nuanced analyses of the spatial divisions of labour within the sector itself. Even the geographies of housing and pensions feed into this since so much of our mortgage finance and our pension savings is handled through that sector. Another thematic exercise might be to pick through the book, not just for the geographies themselves, but for the effects of those geographies. Some of these have already been pointed to above, but they range from the way in which the sharpening geographies of inequality may be pulling the nation apart, to the challenge to democracy implicit in a spatial grammar so centred on London, to the market inefficiencies that result from geographical inequality.

One other theme that might be followed through is that of the international positioning of the UK economy. This is addressed explicitly in a number of chapters, but it is hinted at in others too. The internal geography of a nation's economy is also a reflection of the role it plays in the global economy, whether that be 'workshop of the world' or 'global financial centre'. Likewise, looming decisions about the UK's position in global agricultural and food systems and in energy supply will have implications for the internal economic geography of the country itself. And anyway, it is necessary to think carefully about what is meant by 'the UK economy' (a similar point can be made about the claims of 'local production'). What if material inputs come from around the world, if the ownership of the company is based elsewhere, and/or if the labour itself is drawn from other countries? This increasing geographical complexity of economies forces us (or should force us) to think in terms of relations as well as in terms of territories. Which, again, brings us back to issues of power, for social relations are always in one way or another imbued with power – equal or unequal, disabling or empowering. Thus the economic geography of the United Kingdom, the focus of the chapters here, refers to the internal geography of a particular territory, that of the UK, but always framed within an acknowledgement of the relatedness of that territorially based economy with the global economy that lies beyond. On the other hand we might focus on this wider relationality and investigate the pattern of our imports, the impact of UK-owned companies around the world, the wider geographies of migration on which UK-based employers draw (and the effects of this migration on the economies from which the migrants come).

Whether or not we are at another turning point, when our commonsense understandings will be fundamentally re-evaluated, it is clear, including from the contributions to this book, that we face considerable challenges – about, for instance, the sectoral shape of a future economy, about the deepening inequalities that currently disfigure it, about the future sources of energy, to mention just a few. It is also clear that some well-established nostrums must be seriously revised. It seems clear from the analyses here that relying on single sectors, especially the financial sector, can bring enormous problems. It is clear that a reliance on service sectors above all will not solve the problems adumbrated here. It is clear, too, that the assumed division between macro-economic policy and regional policy is wrong. On the one hand so-called national policies have geographically differentiated effects (think of the bail-out of the banks, or changes to levels of taxation). On the other hand, geographical uneven development affects the functioning of the macro-economy. It seems clear too, and finally, that there needs to be a far more serious politics of geography. The patterns documented here are not just unfortunate outcomes; they matter. We need far more political attention to the geographical shape of the nation. And the papers collected here provide material for thinking about that.

Doreen Massey
The Open University

CONTRIBUTORS

John Allen is Professor of Economic Geography at the Faculty of Social Sciences at The Open University. His research interests broadly concern issues of power and spatiality, most recently in relation to state spatiality, as well as in relation to city networks and finance. He is currently engaged on an ESRC-funded project entitled Biosecurity Borderlands, and is curious about the nature and meaning of biopower. His most recent authored book is *Lost Geographies of Power* (Blackwell, 2003).

Michael J. Bradshaw is Professor of Human Geography at the University of Leicester. His research is on resource geography with a focus on the economic geography of Russia and global energy security. He is co-editor with Philip Hanson, of *Regional Economic Change in Russia,* and with Alison Stenning, *The Post Socialist Economies of East Central Europe and The Former Soviet Union.* Most recently, his teaching and research has focused on energy-related issues and he is currently working on a book entitled *Global Energy Dilemmas*, to be published by Polity Press in early 2012. He is Editor-in-Chief of Wiley-Blackwell's *Geography Compass.*

Gordon L. Clark is the Halford Mackinder Professor of Geography at Oxford University, holds a Fellowship at St Peter's College, and is Sir Louis Matheson Visiting Professor at Monash University's (Australia) Faculty of Business and Economics. An economic geographer with an abiding interest in the tension between global financial integration and national and regional systems, his research focuses upon global finance, governance, and behaviour. Recent books include *Managing Financial Risks: Global to the Local* (edited with Ashby Monk and Adam Dixon, Oxford University Press, 2009), *The Geography of Finance* (with Dariusz Wójcik) (OUP, 2007), *European Pensions and Global Finance* (OUP, 2003) and *Pension Fund Capitalism* (OUP, 2000).

Neil M. Coe is a Reader in Economic Geography at the University of Manchester. His research interests are in the areas of global production networks and local economic development; the geographies of local and transnational labour markets; the geographies of innovation; and institutional and network approaches to economic development. He has published widely on these topics, and is a co-author of *Spaces of Work: Global Capitalism and the Geographies of Labour* (Sage, 2004) and *Economic Geography: A Contemporary Introduction* (Blackwell, 2007).

Kavita Datta is a Senior Lecturer in Human Geography at Queen Mary, University of London. Her research interests focus upon migration, development, finance and gender. Her most recent research has explored financial exclusion among low-paid migrant workers in London and the the role of low-paid migrants in the London economy.

Danny Dorling is Professor of Human Geography, Department of Geography, University of Sheffield. In Britain he has served on the Prime Minister's Academic Advisory Group on Social Mobility. With colleagues he is author of 25 books and 400 papers and reports concerning inequalities in health and other aspects of society. Since 2006, Danny has been working with many others on remapping inequality worldwide (www.worldmapper.org).

Yara Evans is a Visiting Research Fellow in the Department of Geography, Queen Mary, University of London. In recent years she has been involved in research that investigates the lives and work experiences of low-paid immigrants in London, with a particular interest in Brazilians.

James Faulconbridge is a Lecturer in Economic Geography at Lancaster University. His work examines a range of issues relating to globalisation and professional and financial service firms. Recent research has been funded by organisations including the Economic and Social Research Council, the British Academy and the Sloan Foundation with articles being published in journals including *Economic Geography*, the *Journal of Economic Geography* and *Geoforum*. A forthcoming book to be published by Routledge examines the globalisation of advertising agencies in the twenty-first century.

Shaun French is Lecturer in Economic Geography, University of Nottingham. His research interests are in the geographies of financial technologies of risk, financial subjects, and financial exclusion. Current projects include a study of new modalities of life assurance, and the geographies of the buy-to-let market in the UK. He quite often wears white Adidas trainers, even though he is now far too old to carry them off. Recent work has been published in the journals *Environment and Planning D: Society and Space* and *Antipode*.

Chris Hamnett is Professor of Geography at King's College London. He is a well-known writer on various aspects of housing and urban social change and has published widely in the major journals. His most recent book is *Unequal City: London in the Global Arena* (Routledge, 2003). He is currently working with Tim Butler on a book on social class change, ethnicity and education in London, to be published by Polity Press in 2010. He was research director of the Nugee Committee of Inquiry on problems of freehold and leasehold flats which led to the Landlord and Tenant Act, 1987 and has recently been a member of the City of Westminster Housing Commission.

Joanna Herbert is a Visiting Research Fellow in the Department of Geography at Queen Mary, University of London. Her research focuses on migration, specifically using oral histories to explore the experiences of migrant groups and urban inter-ethnic relations. This has included low-paid migrants in London and the South Asian diaspora, with her most recent project exploring the experiences of Ugandan Asians in London and Leicester.

Ray Hudson holds the degrees of BA, PhD and DSc from Bristol University and an Honorary DSc from Roskilde University. Currently Professor of Geography and Pro-Vice-Chancellor at Durham University, his main research interests are in geographies of economies and their relation to territorial development, regional growth and decline. Recent publications include *Producing Places* (Guilford, 2001), *Placing the Social Economy* (with Amin and Cameron, Routledge, 2002) and *Economic Geographies* (Sage, 2005). In recognition of his research, he was awarded the Victoria Medal by the Royal Geographical Society and is a Fellow of the British Academy and of Academia Europeaea.

Brian Ilbery is Professor of Rural Studies in the Countryside and Community Research Institute, University of Gloucestershire. He has research interests in geographical aspects of agri-food systems and local food networks, and has worked on a number of food-related research programmes including a current Rural Economy and Land Use project on plant diseases and food chains and a recent Defra project on organic marketing channels. Brian has published widely on alternative food supply chains, agricultural restructuring and agricultural property relations, including a jointly authored chapter with Damian Maye (2008) on 'Changing geographies of food consumption and production'.

Andrew Jones is a Reader in Economic Geography at Birkbeck, University of London. His research has long focused on globalisation, with a particular focus on transnational financial and business service firms; the globalisation of work; transnational labour mobility and corporate geographies. He has published a wide range of research papers on these topics, and is author of *Management Consultancy and Banking in an Era of Globalization* (Palgrave Macmillan, 2003), *Dictionary of Globalization* (Polity, 2006) and, most recently, *Globalization: Key Thinkers* (Polity, 2010).

Karen Lai is Killam postdoctoral research fellow at the Department of Geography, University of British Columbia. Her research focuses on geographies of money and finance, global cities, markets, capitalism and knowledge networks. She has published in *Environment & Planning A* and *Geoforum*, and is co-author of *Changing Landscapes of Singapore* (McGraw-Hill, 2003). Current research examines the circulation of market ideas and practices in the financial centres of Hong Kong, Shanghai and Beijing, and the business strategies of Canadian banks in Greater China. She received her Bachelor and Master degrees from the National University of Singapore and PhD from the University of Nottingham.

Andrew Leyshon is Professor of Economic Geography, University of Nottingham. His research is in three main areas: money and finance, alternative financial spaces, and the musical economy. Current projects include the geographies of the buy-to-let market in the UK, and the impact of digital technology on the UK music industry. He is the (co-)author or (co-)editor of six books and over 80 academic journal papers and book chapters. He is a member of the Steering Committee of the Financial Services Research Forum, Nottingham University Business School. In 2007 he was elected as an Academician of the Academy of Social Sciences.

Ron Martin is Professor of Economic Geography in the University of Cambridge, and a Professorial Fellow of St Catharine's College there. His research interests include regional growth, evolutionary economic geography, economic theory, regional competitiveness, and the geographies of finance. He has published some 35 books and monographs, and more than 160 papers on these and related themes. He has undertaken numerous research projects for the European Commission. Ron is an editor on the *Cambridge Journal of Economics*, *Economic Geography*, the *Cambridge Journal of Regions, Economy and Society*, and the *Journal of Economic Geography*. He is a Fellow of the British Academy, an Academician of the Academy of Social Sciences, and was selected in 2009 as the Roepke Lecturer in Economic Geography.

Jon May is Professor of Geography at Queen Mary, University of London. His research interests straddle two main areas, with work examining the geographies of homelessness and welfare restructuring, and the politics of migrant labour. Recent research has focused on the role and experiences of migrant workers in low-paid employment in London.

Damian Maye is a Senior Research Fellow at the Countryside and Community Research Institute, University of Gloucestershire. A rural geographer by training, he has research interests in various aspects of agri-food restructuring and rural development. He has published a number of papers on alternative and local food networks, farm tenancy and property relations, and was lead editor on *Alternative Food Geographies* (Elsevier, 2007). He is currently working on a Rural Economy and Land Use project examining plant diseases and food supply.

Cathy McIlwaine is Reader in Human Geography at Queen Mary, University of London. Her research is rooted in geographies of development in relation to poverty, gender and urban violence as well as the nature of North–South linkages through the movement of migrants, especially those located in the lower end of the labour market. Her recent research has focused on the Latin American community in London in relation to coping mechanisms, irregularity and changing gender ideologies among migrants.

Steve Musson is a Lecturer in Human Geography at the University of Reading. His research interests include devolution, local and regional government and the geography of government finance. His research focuses on public–private partnerships in the UK. Previous publications include those in *Environment and Planning A*, *New Political Economy* and *Regional Studies* as well as in a wide range of edited volumes.

Andy Pike is Professor of Local and Regional Development in the Centre for Urban and Regional Development Studies (CURDS), Newcastle University. His research interests are in the geographical political economy of local and regional development. He is widely published in international journals and co-author of *Local and Regional Development* (Routledge, 2006). He has undertaken research projects for the OECD, European Commission, national, regional and local organisations. Andy is current working on brands and branding geographies and devolution, spatial economic policy and economic performance. He leads the Postgraduate Local and Regional Development programmes in CURDS.

Alison Stenning is Reader in Economic and Social Geography in the School of Geography, Politics and Sociology at Newcastle University. She has worked on the economic and social geographies of post-socialism for more than 15 years, focusing particularly on issues of economy, work, class and community. She has published two edited books and a number of book chapters and articles in this field, based on research funded by, amongst others, the ESRC and the Nuffield Foundation. She has recently co-authored *Domesticating Neoliberalism: Spaces of Economic Practice and Social Reproduction in Post-Socialist Cities* (Wiley-Blackwell, 2010).

Kendra Strauss is a researcher with the Inter-University Research Centre on Globalization and Work (www.crimt.org) and part of the Canadian-based SSHRC-MCRI project Rethinking Institutions for Work and Employment in a Global Era. Her current work is on the rise of labour intermediaries and the regulation of temporary employment in the UK. Her other research interests include pension systems and labour market restructuring, with a focus on feminist political economy approaches to gendered economic inequality. Dr Strauss is an Honorary Research Fellow at the University of Glasgow and a Visiting Research Associate at the University of Oxford.

John Tomaney is Professor of Regional Governance and Director of the Centre for Urban and Regional Development Studies (CURDS), Newcastle University, and Professor of Regional Studies, Institute for Regional Studies, Monash University, Australia. His research focuses upon the relationship between territory, democracy, identity and justice, especially at the local and regional scales. He is widely published in international journals and co-author of *Local and Regional Development* (Routledge, 2006). He is also Associate Director of the UK Spatial Economics Research Centre (SERC) and is an Academician of the Academy of Social Science (UK).

Kevin Ward is Professor in Human Geography at the University of Manchester. His research interests are two-fold: first, the changing geographies of the state, in particular with reference to urban and regional governance and economic and social well-being; and second, the changing geographies of work and employment, in particular with reference to the relationships between globalisation, economic and social restructuring and the reorganisation of labour markets. He has written journal articles, book chapters and books on these issues.

Jane Wills is Professor of Geography at Queen Mary, University of London. Her research is in the area of political economy with a particular focus on urban labour markets, labour politics and new forms of political organisation. Recent research has focused on the role and experiences of migrant workers in low-paid employment in London and on London Citizens' living wage campaign.

Neil Wrigley is Professor of Geography, University of Southampton and Editor of the *Journal of Economic Geography*. His research concentrates on economic geography, with a distinctive focus on retail and consumption. His books *Retailing, Consumption & Capital* (Addison-Wesley, 1996) and *Reading Retail: A Geographical Perspective on Retailing & Consumption Spaces* (Arnold, 2002) with Michelle Lowe helped redefine the field, and are complemented by many widely cited research papers covering issues of retail restructuring, competition, regulation and globalisation. In 2004 he was awarded the Ashby Prize, and in 2008 the Royal Geographical Society's Murchison Award, for his publications on these topics.

PART 1

SETTING THE SCENE: UNEVEN ECONOMIC GEOGRAPHIES

PART I

1

INTRODUCTION: THE SHIFTING GEOGRAPHIES OF THE UK ECONOMY?

Neil M. Coe and Andrew Jones

AIMS

- To introduce the economic context and intellectual rationale for the book

- To profile the range of processes influencing the geography of the UK economy over the past two decades

- To explain the structure, style and approach of the book

Twenty years on: re-engaging with the geography of the UK economy

As the 1980s drew to a close, the UK faced what appeared at the time to be a sudden and unanticipated economic crisis. After a severe recession between 1979 and 1982, the British economy had staged a spectacular – if partial and uneven – decade of growth. With Greater London and its financial service sector at the forefront, the Thatcher government had presided over the Big Bang in the City, a house price boom and the development of new information and creative industries (e.g. Hamnett, 1989; Hepworth, 1989; Thrift et al., 1987). Britain appeared to have, at least in part, shaken off the ravages of deindustrialisation and high unemployment, albeit at the expense of an ever widening North–South divide (Martin, 1988). But in 1989 all this came to an end as the economy plunged once more into recession. The then chancellor, Nigel Lawson, raised interest rates to almost unprecedentedly high levels in response to an overheated economy and inflation that threatened to run out of control.

Two decades later, and these events have a familiar resonance. By 2009 the UK economy was once again in the midst of a crisis, namely the worst economic downturn since the Second World War. Again, the boom of the 2000s appeared to have ended with the bursting of a bubble economy, this time focused on the banking sectors rather than the consumer credit flavoured boom of the 1980s. After a long Blair–Brown decade of growth and employment, by late 2009 the UK economy had contracted for a full year, unemployment had risen to over 2.4 million, and house prices had fallen 16 per cent since their 2007 peak (Guardian, 2009a, 2009b). However, perhaps most importantly, after initial focus on the financial sector it soon became clear that those hardest hit by the recession were young people, low-skilled workers, those employed in manufacturing or construction and those living outside of south-east England: in other words, the usual victims.

As far as the recent development of the UK economy is concerned, it seems, therefore, that many of the major trends continue to be cases of history repeating itself. Yet it is also clear that important things have also changed in the last 20 years. In 1989, globalization was a word barely on anyone's lips, whereas now it is perhaps the major force that will shape the UK's economic future. Equally the challenges posed by the need for developing a low carbon economy, along with the associated opportunities for growth, are a novel and unpredictable feature of the contemporary global economic landscape that the UK must engage with. The time for a book that re-engages with some of the longstanding questions about the UK's economic geography in this changed world is thus ripe. The global economic downturn that began in 2007 posed a set of serious challenges for the UK economy, particularly given that the UK appeared to have been more badly affected than either other major European economies like France and Germany or leading Asian economies like China and even Japan. If the UK is to retain its position as one of the leading economies globally, then there is an urgent need to understand the current strengths and weaknesses in a wider global economic context that has changed significantly over the last two decades.

A second rationale for this book, however, is to plug a gap in the economic geographical literature that has emerged in the last 20 years. Economic geographers appear to have been diverted from the political economic approaches of the 1980s that produced a string of powerful contributions concerned with the uneven development of the UK economy and both its social and economic implications. Foremost amongst these contributions were perhaps John Allen and Doreen Massey's edited book *The Economy in Question* (1988) and the associated Reader, *Uneven Re-development: Cities and Regions in Transition* (Massey and Allen, 1988). Written to support an Open University course, these textbooks sought to examine and explain the key questions of uneven development across the UK that had developed over the preceding 20 years since the economic turbulence of the early 1970s. At the time of publication, they represented influential contributions to a vibrant body of economic geographical literature concerned with the UK including books such as Hudson and Williams *Divided Britain* (1989), Lewis and Townsend's *The North–South*

Divide (1989) and Martin and Rowthorn's *The Geography of De-Industrialisation* (1986). Our objective in this book is thus to reconsider the kinds of questions that *The Economy in Question, Uneven Re-development* and this political economic strand of economic geography addressed 20 years ago from the viewpoint of an economic geography that has arguably widened both its theoretical perspective and its understanding of what constitutes the economic realm. Next, we start to profile some of this economic and intellectual background in more detail, before moving on to outline the organisation and approach of the book.

Rethinking the economic geographies of the UK

We should start by explaining what we mean by the term *the economic geography of the UK*. Our use of the term in this book is twofold. On the one hand, it refers to the distribution of economic activities of different kinds across the UK and the underlying processes that produce those uneven patterns. On the other hand, it denotes another meaning – perhaps better thought of as *the UK's economic geography* – which concerns the complex range of multi-scalar connections that constitute the UK economy. These relationships are not neutral but are inevitably about the exercise of power, control and dependency. This, then, is not the UK economy as a bounded space, but as a porous and open system that plugs into economic processes at other scales – most notably within the European Union and global economy (cf. Allen et al., 1998). In other words, it is about the UK's position with the global economic system.

Both dimensions have changed significantly over the past two decades, as subsequent chapters will show. By way of introduction, we would point to six interconnected and overlapping sets of processes that have shaped the UK's space economy in recent times. The list is not exhaustive, but does, we believe, capture the major dynamics at work. We can only give an initial introduction to these processes here – they will be discussed and evaluated in much more detail in the chapters that follow:

- *Globalisation*: The term globalisation captures the increasing functional integration of the global economic system, manifest in the intense but highly geographically uneven cross-border flows of goods, services, money, technology, knowledge and people between national economies (Dicken, 2007). As *The Economist* (2007: 4) describes, 'Britain's economy is one of the most open among the big rich countries'. Trade flows are a good indicator of such openness: in terms of merchandise trade, in 2007 the UK was still the eighth largest exporter in the world (US$438 billion) and the fifth most important importer (US$620bn). The resulting negative trade in goods of some US$180 billion is offset by a strong performance in commercial services trade; exports of US$273 billion in 2007, the second highest in the world, significantly exceeded imports of US$194bn, the third highest figure (www.wto.org, accessed 8/9/09). Flows of foreign direct investment (FDI) in and out of the UK are also extremely high, peaking at US$224bn and US$266bn in 2007 (UNCTAD, 2008).

- *Financialisation*: Central to the UK's integration into the global economic system has been the ongoing financialisation of the economy. This can be seen in two senses: first the sheer size and significance of the financial sector; but second, in terms of the wider influence that it has over the rest of the UK economy. In 2007 the UK financial services industry provided employment for some 1 million workers and generated net exports of US$67 billion – the figure for the US by contrast was US$7bn – helping to offset the negative balance of payments on merchandise trade described above (IFSL, 2009a). In 2008, the UK – with the City of London by far the leading centre – accounted for 70 per cent of the world's international bond transactions, 43 per cent of the over-the-counter derivative market, 35 per cent of foreign exchange turnover, 18 per cent of hedge fund assets and 18 per cent of cross-border bank lending. Another salutary measure of London's influence is that in October 2008, London was hosting US$1700 billion of foreign exchange trading *every day* (IFSL, 2009b).

- *Tertiarisation*: The UK's development into a service economy continues apace. The rapid deindustrialisation that was identified in many studies during the 1980s has persisted: while in 1990 there were some 4.6 million manufacturing workers in the UK – already a dramatic fall from the 7 million of 1979 – by 2009 the figure had dropped to just 2.9 million (ONS, 2009). As a result, the shift to service employment or tertiarization of the UK workforce has also continued: while in 1989 services accounted for 70 per cent of total jobs, by 2000 the figure was 77 per cent, and by 2008, 81 per cent. In 2008, one-quarter of the service employment, or 21 per cent of the overall total, was accounted for by financial and business services, the fastest growing broad employment category in recent times (www.statistics. gov.uk, accessed 8/9/09).

- *Flexibilisation:* The UK labour market has for a long time now exhibited high levels of part-time and temporary working. In June 2009, there were 7.57 million part-time workers in the UK economy, or 26 per cent of the total – of these 1.89 million were men and 5.68 million were women. These figures intersect with 1.45 million workers, or 5.8 per cent of the total workforce, on temporary contracts. All these figures are very high when compared to other leading economies, particularly the proportion of temporary workers, and have slowly but steadily increased over the past two decades (ONS, 2009). While the relative flexibility of the labour market is argued to be attractive to businesses and inward investors, for a significant proportion of the workforce insecure employment has become the norm.

- *Immigration*: Although the UK has long been an important destination for migrants, immigration has become increasingly significant in economic and political terms. While in 1990 the number of international migrants in the UK was 3.7 million (6.5 per cent of the total population), by 2005 the figure was 5.8 million (9.7 per cent) and was predicted to rise to 6.5 million (10.4 per cent) by 2010 (http://esa.un.org/migration accessed 8/9/09). In terms of the workforce, by mid-2009 there were 3.7 million non-UK born workers in employment, accounting for 13 per cent of total employment (ONS, 2009). The accession of several East Central European countries to the EU in 2004 initiated an especially rapid and large wave of immigration of perhaps 1 million workers, with the majority coming from Poland on a short-term basis. The influx triggered a wide range of often highly-politicised debates about the social and economic costs and benefits of immigration, although the economic crisis lead to many European migrants returning home from mid-2007 onwards.

- *Neoliberalisation*: Underpinning the above processes has been a national political agenda that can broadly be construed as neoliberal, a path that was first chartered by the Thatcher administrations in 1980s and continued – to varying degrees – by subsequent Conservative and New Labour governments. Neoliberalisation is used as shorthand for a complex combination of policy trajectories including trade and financial market liberalisation, the privatisation of state assets, the promulgation of flexible labour markets and a scaling-back of direct state influence over economy and society (for more, see Tickell and Peck, 2003). Importantly, these processes should not always be taken as reducing the size of the state apparatus. While that was certainly true under Conservative rule in the 1990s – public sector employment fell from 23.1 per cent of total employment to a low of 19.2 per cent over the years 1992–99 – under New Labour public sector employment has expanded again somewhat, reaching a total of 20.7 per cent of the total, or 6 million jobs, by March 2009 (ONS, 2009).

The economic geographies resulting from the intersection of these powerful forces are many and varied, and can be analysed at different spatial scales. However, what is certain is that they have reworked and yet maintained the powerful structural and spatial inequalities within the UK economy that go under the broad sobriquet of the North–South divide (Amin et al., 2003). Equally certain is that the UK economy is now far more open to the vagaries of the global economy, and the financial system in particular, as was illustrated starkly by the speed with which bad debt in the US housing market precipitated the start of a deep recession in the UK economy in 2007–08.

The approach and organisation of the book

In this final section, we introduce the approach and structure that underpins this volume. The book that follows is not a factual compendium on the UK economy and its position in the world system. Neither is it a systematic and comprehensive review of all research into the UK economy over the past two decades, or, indeed, an intellectual history of the different theoretical perspectives that have been adopted in that research. Nor can we hope to cover every single aspect of the UK economy. Instead, what follows is a series of thematic windows onto the geography of the UK economy, covering the various aspects that we deem to be most pertinent and timely. Each chapter takes a particular topic and seeks to reveal the underlying geographical processes that are at work at various spatial scales, be it intense social networks within the City of London, patterns of uneven regional development within the national system, or the UK's shifting position in global divisions of labour, migrant flows and patterns of trade and investment. The choice of topics deliberately represents a broad take on what constitutes the UK economy in line with the widening sensibilities of economic geography as a sub-discipline (e.g. see Lee and Wills, 1997). Chapters on government finance, housing, pensions and energy, for example, might not have found their way into some economic geography textbooks. Equally, we have unavoidably had to miss other topics out; the public sector, the creative industries, tourism and leisure, and transportation come readily to mind, for example, as topics that could merit inclusion.

It is important to recognise that at the time the chapters were being prepared – 2009 – the UK was in the midst of a severe recession, the final implications of which were still being worked through and debated. Nearly all the authors have touched upon the recession in their analyses and profiled its initial impacts, and several have speculated on the likely outcomes for their particular topic. More considered and detailed analysis of the economic-geographical impacts of the downturn will have to wait for future volumes, however. The chapters themselves are designed to be highly readable and accessible accounts for an undergraduate audience both in geography and beyond, and are deliberately self-contained so they can be used selectively by lecturers, in a different order to that presented here, or to underpin the syllabus of an entire course. In short, we have placed the emphasis on concise, punchy and engaging chapters rather than dense accounts riddled with references and intellectual background. Instead, advice on further readings is designed to offer a way into that background and broader debates for those who want to learn more.

With these various points in mind, the four parts of the book unfold as follows. There are two further chapters in Part I, *Setting the Scene: Uneven Economic Geographies*. Danny Dorling (Chapter 2) provides a multi-variable analysis of the fortunes of England's urban areas, demonstrating that the North–South divide is still very much alive and well. Provocatively, he moves from asserting its continued importance to actually drawing a line on the map to distinguish the two territories. In turn, Ron Martin (Chapter 3) explores the persistent uneven regional geographies underpinning the service-industry-lead 'long boom' of the late 1990s and early 2000s. His analysis suggests that the recent economic crisis is unlikely to change the underlying regional trajectories. Overall, these two chapters offer powerful mappings of the UK's economy and society, thereby serving as invaluable backdrops to the chapters that follow.

Part II, entitled *Landscapes of Power, Inequality and Finance,* contains six chapters that deal with some of the most powerful forces shaping the UK economy over the past two decades, and in particular the nexus of finance and the state that came so prominently to the fore in the financial crisis of 2008–9, as billions of pounds of taxpayers' money were pumped into ailing banks to preserve the integrity of the whole national economic system. While John Allen (Chapter 4) focuses primarily on the City of London, and how we might conceptualise the extraordinary power it exerts over the whole UK economy, Shaun French, Karen Lai and Andrew Leyshon (Chapter 5) consider the UK's financial services industry as a whole and profile other cities – such as Bristol, Edinburgh, Leeds and Manchester – that have also emerged as significant centres of financial employment, although the City of London clearly remains the apex of the system. The following two chapters move the focus onto the state and its attempts to drive economic development in the UK. Steve Musson (Chapter 6) analyses the powerful geographies that underpin government spending of over £600 billion per year in the UK and also profiles the ways in which state funds are increasingly dispersed through forms of public–private partnership. Andy Pike and John Tomaney (Chapter 7) augment this national analysis with a consideration of the impacts of regional devolution on economic development patterns in the UK. The final two chapters in this Part look at topics at the very intersection of state and financial sector activity. First, Chris Hamnett (Chapter 8) debates the multi-faceted position of the housing market within the wider economy and its central role in the financial crisis that started to unroll in 2007. Second, Kendra Strauss and Gordon Clark (Chapter 9) offer a multi-scalar geographical analysis of the challenges facing the UK's pension system, a critical component of the wider financial system.

Part III of the book, *Landscapes of Production and Circulation,* moves the focus beyond the finance-state nexus to examine other key sectoral components of the UK economy. Ray Hudson (Chapter 10) details the profound challenges that manufacturing activities have faced in recent times, showing how these have been heavily shaped by both state policies and shifting international divisions of labour. James Faulconbridge (Chapter 11) charts the inexorable rise of business service activity in the UK economy since the 1980s, a highly geographically uneven growth story that has reinforced the dominance of London and the South East within the wider space economy. Brian Ilbery and Damian Maye (Chapter 12) relate the changing nature of agriculture and food networks in the UK, and in particular the restructuring of arable farming in the past two decades in response to a number of health, environmental and ethical issues. Neil Wrigley (Chapter 13) explores the shifting geographies of UK food retailing against a backdrop of increasing concentration of retail capital, tightening regulation and greater engagement with the global economy, while to conclude the section Michael Bradshaw (Chapter 14) considers the energy challenges facing the UK in relation to issues of energy security and the pressing need to curb carbon emissions.

The final part, *Landscapes of Social Change,* moves the lens onto the labour market and the geographically uneven experiences of living and working in the UK economy. To open, Kevin Ward (Chapter 15) profiles the emergence of a UK labour market that is increasingly characterised by flexibility and insecurity for many workers. Jane Wills and her co-authors (Chapter 16) explore the powerful intersections of these trends with new patterns of in-migration that are producing a new migrant division of labour, most powerfully in London's labour market, but also in other urban areas across the UK. Alison Stenning (Chapter 17) argues that the UK economy has been powerfully affected by 20 years of post-socialism in East Central Europe through a range of economic processes, most notably but by no means only large-scale in-migration since the accession of several East Central European states to the EU in 2004. The volume ends with a short *coda* (Chapter 18) in which we seek to distil some of the key threads and implications of the preceding chapters.

What emerges from the chapters is a series of compelling geographic themes, both old and new. For example, while the formation of a North–South divide through the spatially concentrated decline of manufacturing and concomitant rise of financial and business services was already ongoing by the late 1980s, the UK's labour markets, migrant flows, retail sector, food systems and energy dilemmas, among many others, look markedly different to those that confronted scholars two decades ago. This book is concerned with starting to make sense of these *geographical* stories of continuity and change.

Further reading

- Allen and Massey (1988) and Massey and Allen (1988) still offer powerful tools for understanding the UK economy and are well worth revisiting.
- For positioning the UK against broader economic trends, Dicken (2007) is by far-and-away the best guide to the evolving geographies of the global economy. For a general introduction to economic geography, see Coe et al. (2007).
- For general texts on the human geography of the UK published in the last decade or so, see Dorling (2005), Hardill et al. (2001), Gardiner and Matthews (1999) and Mohan (1999).

References

Allen, J. and Massey, D. (eds) (1988) *The Economy in Question.* London: Sage.
Allen, J., Massey, D. and Cochrane, A. (1998) *Rethinking the Region.* London: Routledge.

Amin, A., Massey, D. and Thrift, N. (2003) *Decentering the Nation: A Radical Approach to Regional Inequality.* London: Catalyst.

Coe, N.M., Kelly, P.F. and Yeung, H.W-C. (2007) *Economic Geography: A Contemporary Introduction.* Oxford: Blackwell.

Dicken, P. (2007) *Global Shift: Mapping the Changing Contours of the World Economy,* 5th edn. London: Sage.

Dorling, D. (2005) *Human Geography of the UK.* London: Sage.

Economist, The (2007) Britannia Redux: Special Report on Britain, 1 February, p.4.

Gardiner, V. and Matthews, H. (eds) (1999) *The Changing Geography of the United Kingdom,* 3rd edn. London: Routledge.

Guardian, The (2009a) Rising UK Unemployment, 11 November.

Guardian, The (2009b) Nationwide doubts housing crash is over, despite fifth month of rising prices, 2 October.

Hamnett, C. (1989) The political geography of housing in contemporary Britain, in J. Mohan (ed.), *The Political Geography of Contemporary Britain.* Basingstoke: Macmillan.

Hardill, I., Kofman, E. and Graham, D. (2001) *Human Geography of the UK: An Introduction.* London: Routledge.

Hepworth, M. (1989) *Geography of the Information Economy.* London: Belhaven.

Hudson, R. and Williams, A.M. (1989) *Divided Britain.* London: Belhaven.

International Financial Services, London (IFSL) (2009a) *IFSL Research: Foreign Exchange 2009.* London: IFSL.

International Financial Services, London (IFSL) (2009b) *IFSL Research: International Financial Markets in the UK.* London: IFSL.

Lee, R. and Wills, J. (1997) *Geographies of Economies.* London: Arnold.

Lewis, J. and Townsend, A.R. (eds) (1989) *The North–South Divide.* London: Paul Chapman.

Martin, R. (1988) The political economy of Britain's North–South divide, *Transactions of the Institute of British Geographers,* 13: 389–418.

Martin, R. and Rowthorn, B. (eds) (1986) *The Geography of De-Industrialisation.* London and Basingstoke: Macmillan.

Massey, D. and Allen, J. (eds) (1988) *Uneven Re-development: Cities and Regions in Transition.* London: Hodder and Stoughton.

Mohan, J. (1999) *A United Kingdom? Economic, Social and Political Geographies.* London: Edward Arnold.

Office for National Statistics (ONS) (2009) *Labour Market Statistics (August).* London: ONS.

Thrift, N., Leyshon, A. and Daniels, P. (1987) 'Sexy Greedy': The new international financial system, the City of London and the South East of England. Working Papers on Producer Services, University of Bristol and Service Industries Research Centre, Portsmouth Polytechnic.

Tickell, A. and Peck, J. (2003) Making global rules: globalization or neoliberalization? in J. Peck and H.W-C Yeung (eds), *Remaking the Global Economy: Economic-Geographical Perspectives.* London: Sage. pp. 163–81.

UNCTAD (2008) *2008 World Investment Report.* Geneva: UNCTAD.

2

PERSISTENT NORTH–SOUTH DIVIDES

Danny Dorling

AIMS

- To describe the extent of the social, economic and political North–South divide to the human geography of Britain as of 2008

- To give some early indications as to how the divide appeared to be sharpening with the advent of the economic crash of 2007–9

- To suggest that we now know enough to be less vague about defining the North–South divide

- To focus on the English North–South divide because it is worth telling longer and separate stories for the other countries of the Isles

In this chapter I assess the current extent of the North–South divide in England and recent trends in the divided human geography of the country. I will argue that the North–South divide has grown in importance since the early 1970s. England, it seems, is a country which is split in two and increasingly at unease with itself and what it is becoming. But how is the human geography of Britain as a whole now best summarised? With colleagues a few years ago I undertook research sponsored by the government department responsible for English cities to create a database of many aspects of their human geography and how they were changing. This State of the Cities Database (SOCD) (see www.socd.communities.gov.uk) comprised 75 indicators at seven different spatial levels and at different points in time for some of the variables (Parkinson and collective, 2006). In this chapter I will try to give a flavour of the information held in the database by focusing on five themes: life expectancy; poverty; education and skills; employment; and wealth. It can be argued that these themes pertain to traditional measures of quality of life as seen through lack of disease, ignorance, idleness, want and squalor, all reflected through their modern-day equivalents of high life expectancy, good qualifications, low work-related benefit claims, low rates of poverty and reasonable house prices.

It should be noted that four aspects of the database are unique. First, amongst other geographies, it collates data for major cities in England as defined by their built-up urban boundaries. This allows cities to be compared in a way which is not influenced by whether their official administrative boundaries happen to incorporate a great deal of their hinterland or not. Second, the database collects very up-to-date information as well as data from the past to allow comparison using comparable boundaries. This allows changes over time for these consistently defined areas to be calculated and shown. Third, the database spans a very wide range of indicators. This allows many aspects of life in cities to be compared. Fourth, the database, where possible, presents data for over 1,000 'census tracts' within and outside of these cities, which can also be compared over time. So using this data, how can a picture be painted of the state of England's cities as reflected through their populations, and the changes to the fortunes of those populations, over time?

Maps, although out of fashion in much contemporary English geography, are useful here. In this chapter both conventional maps and a Tetris map (population cartogram) of cities are presented. On the conventional map the urban boundary of each city is shown, but many cities of course appear just as specks on the national map. On the Tetris map each city is presented as the collection of tracts which constitute it on a rough population cartogram of the country. The Tetris map is far more useful for visualisation, but it requires a little patience in learning which shapes are which cities (see Figure 2.1).

The chapter is structured to look first at inequalities in life expectancy, then poverty, then in education, employment (just prior to the crash of 2008), and then in wealth (at a similar point in time). A line is suggested where the North–South divide can be said to run, and an argument is made that it is worth drawing the line quite precisely. The chapter concludes to suggest that the divide has deepened in recent years. The initial data coming out after the 2008 crash confirms this, as the North has been most badly hit and least well supported. The 'bail out' was for the South.

Life expectancy – grim up North?

As has been well chronicled, many northern cities in England have felt the full force of deindustrialisation since it took hold in the late 1970s. Manufacturing industry in Britain continues to decline as fast as it ever did – faster even very recently given the downturn that started in 2008 (and see also Chapter 10). But how does this relate to measures such as life expectancy? The people of Stoke, for example, live on average almost 77 years each

Figure 2.1 The Tetris map of cities in England – a key to their location in population space

(see Table 2.1). This does, however, put the city in the second worse of the five groups shaded in Figure 2.2, which shows the spatial distribution of life expectancy by city from birth for the years 2001 to 2003. Driving from the South through Birmingham to Stoke means driving past people who on average are destined to live two or three fewer years less than the highest averages of almost 80, four years down by the time you hit the cities of the north–west. This is an old pattern of inequality, but one which has strengthened in recent

Table 2.1 Key state of the city indicators, sorted by an overall score (and change measure provided in final column) divided into six leagues

Division	City	Life exp. 2001–2003	2001 per cent of adults with a degree	per cent working age claiming JSA/IS 2003	Percent of poverty by PSE 1999–2001	Average housing price 2003	Average score 2003	Change in score over time
Premiership	Cambridge	79.5	41	5.1	29	244862	82.3	5.6
	Aldershot	79.0	22	3.7	17	238991	81.9	4.6
	Reading	79.6	26	4.7	20	211794	81.5	5.3
	Oxford	79.2	37	6.1	30	255181	80.9	6.7
1st	Crawley	79.6	19	4.8	22	205506	79.8	4.7
	Bournemouth	79.7	17	7.1	21	214296	79.1	5.0
	York*	79.4	23	5.4	25	147513	78.2	5.4
	Worthing	78.8	16	6.4	20	186992	78.0	4.1
	Brighton	78.4	29	9.3	27	212361	77.6	6.8
	Southend	79.0	13	7.5	19	186481	77.6	4.2
	London	78.6	30	10.3	33	283387	77.5	6.7
	Bristol	78.9	23	7.7	25	160708	77.1	4.6
	Southampton	78.8	19	6.9	25	172585	76.9	4.9
	Norwich	79.8	18	7.5	27	138187	76.3	3.9
	Portsmouth	78.8	16	6.6	25	157145	76.2	4.0
	Milton Keynes	78.2	18	6.6	25	161625	76.0	4.2
	Swindon	78.2	15	6.6	22	150689	76.0	4.1
	Gloucester	78.4	16	8.5	22	141690	75.5	3.4
	Warrington*	77.9	17	6.8	23	119668	75.1	4.6
	Northampton	78.2	17	7.8	24	135871	75.1	4.1
	Ipswich	79.0	16	10.1	25	134514	74.7	3.5
	Chatham	77.7	12	7.7	23	142374	74.2	2.8
2nd	Preston*	77.7	17	7.2	26	97038	73.6	3.6
	Derby*	78.1	18	10.5	27	114280	73.1	4.2
	Leeds*	78.2	19	8.9	32	119262	72.8	3.1
	Nottingham	77.5	18	9.8	28	123663	72.7	3.4
	Telford*	77.9	13	9	27	115722	72.6	2.9
	Leicester	78.0	17	11	28	124812	72.6	2.3
	Blackpool*	77.2	13	8.9	24	103656	72.5	2.2

(Continued)

Table 2.1 (Continued)

Division	City	Life exp. 2001–2003	2001 per cent of adults with a degree	per cent working age claiming JSA/IS 2003	Percent of poverty by PSE 1999–2001	Average housing price 2003	Average score 2003	Change in score over time
	Plymouth	78.1	13	9.8	28	118978	72.4	3.5
	Hastings	77.4	15	13.4	25	163128	72.3	3.7
	Luton	77.2	14	9.7	28	143698	72.2	2.0
	Wakefield*	77.5	14	9	28	110407	72.1	3.5
	Peterborough	77.5	14	9.5	28	123089	72.1	2.1
	Coventry*	77.8	16	10.9	28	111165	72.0	3.8
	Huddersfield*	77.2	15	8.7	29	97815	71.6	1.8
3rd	Manchester*	76.7	19	11.6	30	119569	70.9	3.6
	Sheffield*	77.9	16	10.4	33	96328	70.8	3.6
	Wigan*	76.5	12	8.6	27	88946	70.7	2.6
	Birkenhead*	77.9	13	12.2	29	95632	70.6	3.3
	Bolton*	76.8	15	10.4	29	89281	70.4	2.3
	Mansfield*	77.1	9	9.4	28	94749	70.4	2.0
4th	Grimsby*	77.6	10	11.5	28	77898	70.0	2.6
	Doncaster*	77.3	11	10.6	30	82267	69.8	3.0
	Birmingham*	77.4	14	12.8	33	122794	69.7	2.2
	Stoke*	76.9	11	10.3	29	78834	69.7	1.7
	Newcastle*	77.1	16	12.8	34	111220	69.2	4.1
	Barnsley*	77.2	10	10.8	32	79492	68.9	3.0
	Rochdale*	76.4	14	12.2	31	92523	68.8	2.5
	Burnley*	76.8	12	10.7	31	55879	68.7	1.5
	Bradford*	76.9	13	11.5	33	75919	68.6	1.4
	Middlesbrough*	77.1	12	13.1	32	81760	68.4	3.1
5th	Sunderland*	76.6	12	12.4	34	91322	67.8	3.2
	Blackburn*	75.8	14	12.7	30	70969	67.8	1.9
	Hull*	76.6	12	17.1	33	72374	66.0	1.4
6th	Liverpool*	75.7	14	18	36	87607	64.7	2.8

Note: cities in the North of England are marked by an asterix

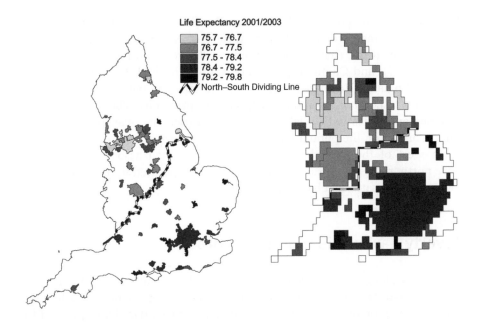

Figure 2.2   Life expectancy from birth 2001–3

decades. The very latest mortality data, for 2006 and 2007, shows inequalities in health across Britain returning to relative levels of inequality last seen in the 1920s and 1930s. At the height of the economic boom, just at the cusp of the crash, the current had returned again to 'brass tacks' inequalities in wealth and health that it had last known in those years and months leading up to the 1929 precursor of 2008 (Dorling and Thomas, 2009).

Figure 2.2 shows very recent estimates, where Local Authority figures for men and women have been aggregated on the basis of the Local Authority populations which best fit the built-up urban areas of each city to produce average life expectancies for all the inhabitants of those cities (both men and women combined). The map shades cities so that those with life expectancies of similar year of age are shaded the same tint. Thus cities are shaded the darkest where residents, on average, currently live for three score years and nineteen (79). The precise calculations used to estimate life expectancy are provided by the country's Office for National Statistics (ONS), and the figures presented on the map above are population-weighted averages of those figures. For these cities, life expectancy in England is highest in Norwich at 79.8 years, and lowest in Liverpool at 75.7 years; there is a clear North-west–South-east gradient to life expectancy.

The only significant anomalies to this gradient in the North are York, with an average life expectancy of 79.4, and Leeds, with 78.2. York sits in a vale of relative affluence in the North of England and so its exception is perhaps of little surprise. The figure for Leeds is partly the result of the Leeds conurbation not being as extensive as, for instance, that of Manchester in population.

Were Leeds to include its neighbour of Bradford with a life expectancy of 76.9, then the map would look quite different. Nevertheless life expectancy tends to rise to the east of the Pennines. The two southern anomalies are Hastings (77.4) and Chatham (77.7), areas also with high rates of poverty for southern England. Hastings and Chatham suffer from particularly bad transport routes to London given their geographical proximity. Them aside, a circle of towns and cities with relatively high life expectancy can be seen to surround London on the population cartogram in Figure 2.2 – broken only to the North–west of the capital where places – too much associated with their more northern neighbours – are not so well incorporated into the centre. But how are these patterns changing?

Poverty – where is it getting worse?

Analysis of changes in measures such as imprisonment rates and child poverty demonstrate a more subtle geography than a simple North–South divide of the more static images. Figure 2.3 shows one particular change, that of the spatial distribution of rises in the rate of poverty by city between 1991 and 2001. Nowhere over this time-period was the rate of poverty recorded as falling when consistently measured.

Because incomes have only been calculated at one point in time by ONS (1998) it is not possible to compare changes over time, especially in income

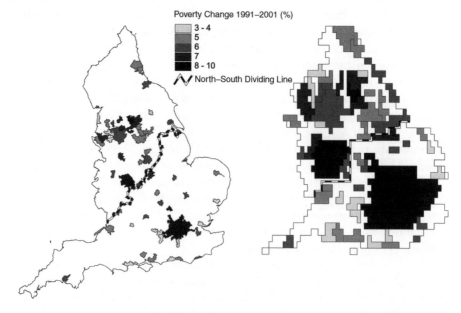

Figure 2.3   Change in rate of poverty by city – initial estimates 1991 to 2001

that has been equivalised for household composition, net of taxes and calculated after taking into accounting housing costs. Furthermore the ONS estimates do not include estimates of the distribution of income in each area, particularly that proportion of the population living below 60 per cent of the medium national income. To allow for an estimate of the changing rates of poverty in each city in the absence of such income data the Figure 2.3 shows the changing proportion of households estimated to be living in poverty according to calculations made following the 1990 Breadline Britain survey and the 1999 Poverty and Social Exclusion Survey (by researchers working with the Joseph Rowntree Foundation). For details of how these figures were calculated for small areas see Dorling and Thomas (2004).

Rates of poverty have increased in all cities since the early 1990s by this measure. That is, a higher proportion of households over time do not have access to the resources that most people think are necessary to live a decent life. Such rates of poverty can and do grow as rates of affluence also rise in cities. The highest increase, of an additional 10 per cent of the population living in poverty over the course of the 1990s, is found in London, followed by a 9 per cent rise in Luton, Birmingham and Bradford. The lowest increases, of an extra 3 per cent of the population living in poverty are found in Aldershot, Swindon, Warrington and York. Estimates of poverty made using techniques similar to these are soon to be incorporated in official government statistics and so this figure gives an impression of how these new statistics should show high and rising rates of poverty even in generally affluent large cities, and especially in the capital, as socio-economic polarisation has risen. One result of both the numbers living in poverty rising and the riches of the wealthy in Britain increasing dramatically is that those in the middle begin to feel quite badly off and increasingly threatened. That sense of insecurity was made far more real when unemployment rapidly rose past two million in early 2009.

Education – where do the skilled travel to?

Educational divides in Britain did not end with the demise of the 11-plus exam. Children are divided at ages 17 and 18, when, according to exam results at these ages, one-third are now drawn to go to university. What is most telling is where they then move on to. For many students from the South, if they have lived in the North it will be for three years spent in places like Durham, or York, Manchester or Leeds, before heading south again upon graduation. Just to make the point clear, Figure 2.4 shows the spatial distribution of the individuals in cities holding a university degree in the year 2003, as a proportion of the total economically active population.

The highest concentrations of the economically active population in England qualified to degree level are observed in Cambridge, Oxford and London, and also York, Warrington, Bristol, Crawley, Norwich and Brighton,

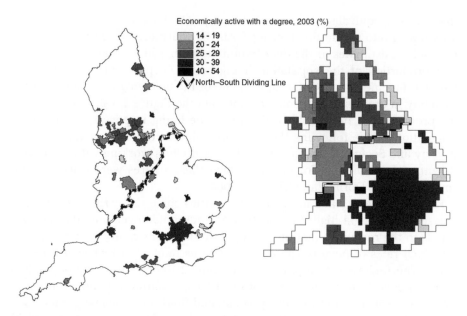

Figure 2.4 Proportion of economically active population with a university degree, 2003

where the proportion of the economically active population with a degree is over 40 per cent. In contrast, the smallest proportions (ranging from 14 per cent to 19 per cent) are observed in Sunderland, Hull, Grimsby, Doncaster, Stoke, Peterborough, Southend and Chatham.

Look at Figure 2.4 again. Now the impression of a clean North–South divide is complicated by a series of colonial outposts. Centred from London they appear at 12 o'clock to the north where York is found; then Norwich at 2 o'clock; Brighton at 6 o'clock; Bristol at 9 o'clock; and Warrington at around 10.30. The country cannot be governed from London alone. Around the periphery outposts are required where those educated to the higher levels can cluster together in safety and mutual understanding up the spokes of their various motorways from the centre. Closer to home, Oxford and Cambridge are both just an hour's drive down newly built 6-or 8-lane roads to the centre of power. However, with a declining manufacturing base in the provinces and increasing reliance on one industry in the capital (finance) it becomes harder to see what all this organisation is for (and see Chapters 3 and 6). For that we need turn to issues of what the English now do: employment.

Employment – and those seeking it

Even in the pre-2008 economic boom times a remarkable number of people of working age in England were unemployed. Their geographical distribution is shown in Figure 2.5. A much higher number cannot work because they are

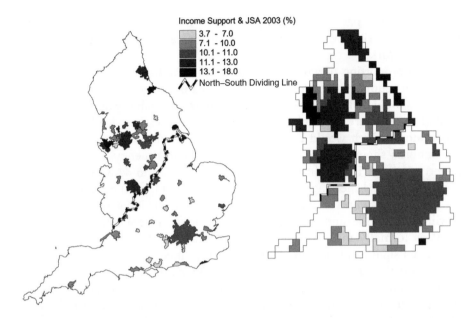

Figure 2.5 Proportion of the working-age population living on Income Support or Job Seekers Allowance (JSA, 2003)

now ill, often suffering from depression. In contrast the geographical distribution of the population living on unemployment benefits is of a constantly changing population. Very few people now live on these benefits for long periods, but many come on and off benefits, repeatedly, through their lives. Figure 2.5 shows the proportions of adults of working age (as estimated by the statistics quango NOMIS) who were claiming Income Support or Job Seekers' Allowance by August 2003.

There are many ways in which lack of work can be measured and many of these are included in the database that this chapter draws on. However, given problems of changing definitions of unemployment over time and of the welfare benefits associated with unemployment and low-paid part-time employment, the combination of the two benefits shown in Figure 2.5 provides one of the longest reliable time series available for small areas.

By the middle of 2003 (the latest date for which my colleagues and I had numerator and denominator data when creating this picture for the British government) some 18 per cent of the working-age population of Liverpool and 17 per cent of that population of Hull were claiming these benefits. The next four cities with the highest claimant rates, all of 13 per cent of their adult populations, were Birmingham, Hastings, Newcastle and Middlesbrough. This partly explains Hastings featuring as an anomaly in the South. Other cities with more than 11 per cent of their working-age populations living on these benefits include: Blackburn, Sunderland, Birkenhead, Rochdale, Manchester, Bradford and Grimsby. The figure for London was 10.3 per cent.

Rates below 6.5 per cent of this population were found only in Aldershot, Reading, Cambridge, Crawley, Oxford, Worthing and our anomalous friend in the North: York.

A high proportion of working-age people in English cities have to rely on benefits to support themselves, mainly because they cannot find suitable work. It should be noted here again that people on disability and other health-related benefits are not included in these maps (nor men aged 60–64 in the numerator), which would both further inflate these numbers and reinforce the patterns shown above. The parliamentary constituency with the highest poverty rates in Britain in recent years recorded up to 41.3 per cent of the population living there aged between ages 25 and 44 relying on benefits, and up to 60.3 per cent of the population aged 45–59 doing so (Thomas and Dorling, 2007: 143 and 178). There had been significant falls in unemployment as formally measured in the years immediately prior to 2003. There have been significant rises in almost all places since 2007. The picture shown here is about as good as it got over the last 30 years.

Many cannot find work, but by 2008 more people than ever were also working in Britain. Both these trends are true because each year there are fewer people in the country who are neither working nor unemployed. Women have been coerced to re-enter the labour market after having children more quickly than the year before. People are encouraged to work longer before retiring, and students to work through their studies: work for longer work longer hours, work harder, and – in practise – work for less. More people in Britain who can afford to try to buy a house do so with far more difficulty than they would have done in recent decades; this despite more of us now working and owning on average more houses per household! We work harder, in greater numbers and for longer, to get by. We are rewarded increasingly unevenly for that work. With the economic crash that hit during 2008 and became so much worse in 2009 it is easy to forget that all was far from well in a country as divided as Britain even before the job market declined. One result is a huge rise in private renting for those whose parents would have taken out a mortgage, with implications for the geography of wealth.

Wealth – and the changing cost of shelter

Figure 2.6 shows the average absolute change in equivalised (for type of home) housing price from 1993 to 2003 using building society records as the source of data for the earlier data, coupled with the 1991 and 2001 census figures on dwelling type. Because the sums of money involved are so large and because cities start off from different bases, it makes more sense to show absolute rather than relative change.

Average housing prices in the ten years 1993–2003 rose by over £200,000 only in London. They rose by more than £150,000 in Oxford, Cambridge, Aldershot, Brighton and Bournemouth; and by more than £100,000 in Reading,

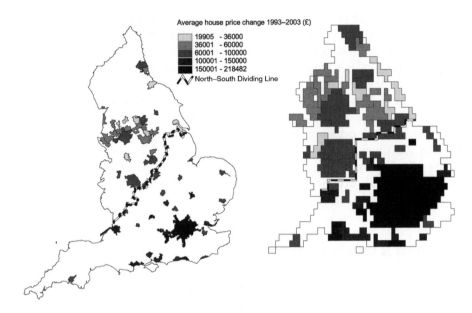

Figure 2.6 Change in average housing price in 1993–2003

Crawley, Worthing, Southend, Milton Keynes, Southampton, Hastings, Bristol and Portsmouth. They rose between only £40,000 and £50,000 in Preston, Sheffield, Birkenhead, Sunderland, Huddersfield, Rochdale, Wigan, Liverpool and Bolton; by between £30,000 and £40,000 in Doncaster, Grimsby, Middlesbrough and Stoke; by between £20,000 and £30,000 in Bradford, Barnsley, Hull and Blackburn; and by just less than £20,000 in Burnley.

Changes in housing prices over time bear very close correspondence to changes in housing wealth. Although the use of the dates 1993 to 2003 above show a period of particularly rapid polarisation in housing prices, that polarisation has been continuing fairly constantly since digital reports of prices were first made (in the early 1980s). Short-term falls in house prices, as occurred in the early 1990s, do little to dent the long-term trend in polarising prices between English cities along, and exacerbating, a North–west to South–east divide (Ridge and Wright, 2008).

Those now living in London are fearful of leaving it as they will not be able to afford to return. Those living in the North cannot move to London until they have no children and so need little space, or have a relative rich enough to finance their move. London, increasingly, only has space for the best and worse-off in Britain (and from abroad). Other than asylum-seekers, the very worse-off are housed still by the state. The very best off are building huge swimming pools under their Westminster and Kensington mansions, palatial home cinemas, underground garages for their multiple cars, and remodelling their interior decorations regularly in those parts of the capital home to the world's super-rich.

As of 2008 these prices began to fall rapidly, but at the time of writing not back to 2003 levels (and see Chapter 8). Prices fell faster in the North where they were not protected by the lack of supply and huge demand for housing which remains in the South. It also became apparent during the early 1990s that rises in negative equity were fastest in the North, where recent buyers had both been less likely to have a deposit and so more likely to have a 100 per cent mortgage, and where the immediate price falls were more acute. As recession/crash/depression hit, the North–South divide described in this chapter swiftly sharpened. During 2008 and 2009 we first began to learn that in the most expensive parts of the South of England housing prices continued to rise where all around them prices were falling – falling fastest further North and West (see Dorling, 2010).

Drawing a line on the map

A useful summary of England's human geography can be created by aggregating the variables described so far to create a very simple overall index of the state of the English cities, combining the five traditional measures of quality of life as seen through lack of disease, ignorance, idleness, want and squalor as reflected through their modern day equivalents of high life expectancy, good qualifications, low work-related benefit claims, low rates of poverty and high house prices. These key state-of-the-city indicators are summarised in Table 2.1. As can be seen, the indicators are sorted by an overall score (and a change-over-time measure is given). This overall index confirms the general impression given by the more than one hundred maps and cartograms contained in the full report on which this chapter draws. The impression is that, in general, English cities are clearly divided between those in the South-east of the country and those situated towards the North-west. And the South-east is increasingly dominated by London.

But where exactly does any dividing line fall? When asked where the North–South divide is, geographers in Britain have a tendency to give vague answers. Although this is understandable, I think we now have enough detailed information on life chances, political views, health and wealth in Britain to be able to say with a little more certainty where the line lies (see Figure 2.7). This is the line that separates upland from lowland Britain, the hills from the most fertile farmland, areas invaded by Vikings from those first colonised by Saxons. Numerous facts of life divide the North from the South – there is a missing year of life expectancy north of this line. Children south of the line are much more likely to attend Russell group universities. For those that do go to University (and they often go to the North to study!), a house price cliff now runs along much of the line, and, on the voting map, the line still often separates red from blue.

By county the North lies above the old counties of Gloucestershire, Warwickshire, Leicestershire and Lincolnshire and 'nips' only into parts of

Figure 2.7   The North–South dividing line

some of those counties. Most of each of those counties, and all the areas of England below them, are in the South. By constituency, the North includes and lies above the new parliamentary constituencies of the Forest of Dean on the north bank of the Severn; includes West and Mid Worcestershire, Redditch, Bromsgrove (and hence all of Birmingham), Meriden, Coventry South and North East, Warwickshire North, Nuneaton, Bosworth, Loughborough, Rushcliffe, Newark, Bassetlaw, Brigg and Goole, Scunthorpe, Cleethorpes, ending at Great Grimsby and the south bank of the Humber. It would be possible to go further and split some of these constituencies in half. It would be possible to identify enclaves and exclaves along the border, but this would suggest too much of a rigid line, and the border does move, especially when a new motorway is built or train line to London improved.

Within the North are places that look and sometimes act (e.g. vote) like the South. Areas around the vale of York and Cheshire are contenders here – but they are still northern. Similarly there are parts of the South, especially within London, that are very unlike much of the rest of the South, but they are still southern. Scotland and Wales are part of the North, despite having managed to eschew the Victorian attempts to label them North and West Britain respectively. In terms of life chances, the only line within another European country that is comparable to England's North–South divide is that which used to separate East and West Germany. This is found not just in terms of relative differences in wealth either side of the line, but most importantly in terms of health, where some of the extremes of Europe are now found within this one divided island of Britain.

Conclusion: deepening divides?

There is little sign of the divide narrowing and many indications that, again in general, it is widening. The same few exceptions to this generalisation have already been mentioned; most notably York ranks within southern cities (7th overall, and the only northern city in the top dozen). All 23 cities at the bottom of Table 2.1 are in the North as defined in the research project this data arose from (by Government Office Region with the West Midlands in the North). The next four are all southern, but those four are socially and/or spatially at the greatest distance from the capital: Hastings, Plymouth, Luton and Peterborough. There is no southern city which, overall on all five indicators, compares badly or even equally to any of the worst-off 20 Northern cities. In general, the better off a city was on these scores in the recent past, the more it had improved in the period to 2003.

A simpler way to put this is to state what it would take for Liverpool, at the bottom of the table, to become like Leeds, midway, and for Leeds to become like Cambridge (at the top). For Liverpool to be like Leeds, its peoples' life expectancy would have to rise by 2.5 years more than that of Leeds in the future, 5 per cent more of its adult population would need to gain a degree, 9 per cent of the working age population would have to come off Income Support or Job Seeker Allowance benefits (and none off such benefits in Leeds), overall poverty would have to fall by 4 per cent and average housing prices rise by £31,650. For Leeds to be like Cambridge, life expectancy in Leeds would have to increase by 1.3 years more than in Cambridge in the near future, an extra 22 per cent of the population would need to gain a degree, 4 per cent fewer people would need to be on work-related benefits, poverty rates would need to fall by 3 per cent and house prices would have to rise by an average of £125,600 per home.

English cities can appear in a series of leagues when the data in Table 2.1 is considered in the round. A 'premier league' of four cities with high average scores from 80.9 to 82.3 is clear (including Oxford and Cambridge), followed by 18 'first division cities' with scores from 74.2 to 79.8 (from Crawley to Chatham, including London and Bristol). There is a gap and then a 'second division' of 14 cities scoring between 71.6 and 73.6 (from Preston to Huddersfield, including Leeds and Nottingham), followed by a 'third division' from 70.9 to 70.4 (headed by Manchester and down to Bolton), and a 'fourth division' from Grimsby to Middlesbrough, including Birmingham and Newcastle); with Blackburn, Sunderland; and then Hull as a fifth; and then Liverpool following below in division six, an English city in a group of its own (if other cities in the United Kingdom were included outside of England, Liverpool might potentially be joined in a group by Swansea, Glasgow, Belfast and other similar western ports and old industrial centres).

Almost all southern cities are in the premier league or first division of Table 2.1. Less than a half dozen are found in the second division, and none below that. Division two downwards is dominated by cities of the North of England. To borrow from the subtitle of a recent atlas of poverty produced for the United States (Glasmeier 2005), England, as viewed through the lens of its cities, is 'one nation, pulling apart'. Not to state this clearly in conclusion would be unfair to the reader as the patterns are so clear. Given how obvious such a conclusion is from the maps reproduced here, it is imperative that this simple truth is not lost in the study of the nuances of more subtle changes occurring in urban England as revealed by this data.

Further reading

- Baker and Billinge (2004) provide a detailed overview of the different dimensions to the English North–South divide and how it has moved over time.
- Haworth and Hart (2007) offer further insights into issues of well-being and in particular the debates around community and inequalities within regions.
- Take a look at Roberts and McMahon (2007) for a wide-ranging analysis of geographical divisions in relation to crime and social justice divides between and within communities.
- For other excellent accounts on social and geographical inequalities in the UK see Elliot and Atkinson (2007), Irvin (2008) and Wilkinson and Pickett (2009).

References

Baker, A. and Billinge, M. (eds) (2004) *Geographies of England: The North–South Divide, Material and Imagined.* Cambridge: Cambridge University Press.

Dorling, D. (2010) *Injustice: Why Social Inequality Persists.* Bristol: Policy Press.

Dorling, D. and Thomas, B. (2004) *People and Places: A 2001 Census Atlas of the UK.* Bristol: Policy Press.

Dorling D. and Thomas, B. (2009) Geographical inequalities in health over the last century, in H. Graham (ed.), *Understanding Health Inequality.* Maidenhead: Open University Press.

Elliot, L. and D. Atkinson (2007) *Fantasy Island: Waking Up to the Incredible Economic, Political and Social Illusions of the Blair Legacy.* London: Constable and Robinson.

Glasmeier, A. (2005) *An Atlas of Poverty in America: One Nation, Pulling Apart, 1960–2003.* Pennsylvania, PA: University of Pennsylvania.

Haworth, J. and Hart, G. (eds) (2007) *Well-Being: Individual, Community, and Social Perspectives.* Basingstoke: Palgrave.

Irvin, G. (2008). *Super Rich: The Rise of Inequality in Britain and the United States.* Cambridge: Polity.

Parkinson, M. and collective (2006) *The State of the English Cities, Volumes 1 and 2.* London: ODPM.

Ridge, T. and Wright, S. (eds) (2008) *Understanding Poverty, Wealth and Inequality: Policies and Prospects.* Bristol: Policy Press.

Roberts, R. and McMahon, W. (eds) (2007) *Social Justice and Criminal Justice.* London: Centre for Crime and Justice Studies.

Thomas, B. and Dorling, D. (2007) *Identity in Britain: A Cradle-to-Grave Atlas.* Bristol: Policy Press.

Wilkinson, R. and Pickett, K. (2009) *The Spirit Level: Why More Equal Societies Almost Always Do Better.* London: Allen Lane.

Acknowledgements

A version of this chapter, and Figures 2.1–2.6, were previously published in the French Geography Journal *Geocarrefour*, in 2008 under the title 'London and the English Desert: the grain of truth in a stereotype' (83(2), 87–98). Figure 2.7 is taken from Figure 7.3 in Dorling (2010).

3

UNEVEN REGIONAL GROWTH: THE GEOGRAPHIES OF BOOM AND BUST UNDER NEW LABOUR

Ron Martin

AIMS

- To show how national economic growth is an inherently geographically uneven process

- To illustrate this through an analysis of what New Labour called the 'longest boom in Britain's history'

- To show how the recession that followed, reckoned by many to be one of the worst of the post-war period, has also proved to be a highly uneven process geographically

Geography and the economy

Explaining economic growth has long been the focus of intense debate within economics. Different theories emphasise different causal mechanisms and processes, and indeed reflect different underlying conceptions of what the economy is (or should be) and how it functions. For these reasons, different theories can lead to quite different policy implications, to different political-ideological positions on how the economy should be managed. Standard neoclassical growth theory attributes the performance of the economy to the quality and efficiency of utilisation of its basic factors of production, labour and capital. It emphasises the importance of free markets and the free mobility of capital and labour in promoting growth, but it has little to say about the role of technological innovation, which is assumed to be 'autonomous' or 'exogenous'. Keynesian economics puts the focus on the determinants of aggregate demand and investment, on the expectations and behaviour of consumers and producers, and argues that if left to its own devices the free market economy is systemically prone to unstable growth, to cycles of boom and bust. New classical economic theory follows much of the standard neo-classical approach, but in addition assigns a key role to managing the money supply and controlling inflation in order to secure free-market based growth. And the so-called new endogenous growth economics likewise takes the standard neoclassical model as its starting point, but seeks to incorporate technological innovation, rather than assuming it to be autonomous (hence the epithet 'endogenous'), and to allow for the impact of increasing returns effects in the use of capital and labour.

What is common to all such theoretical frameworks, however, is a neglect of geography. The 'economy' is considered to exist at essentially two levels – the macro-economy of nation-wide aggregates (overall consumption, investment, exports, imports and the like) and their inter-relationships; and the micro-economy of individual households and firms (or industries). Nowhere in these theories of economic growth is there any explicit recognition that the 'everyday business of economic life' (to use Alfred Marshall's phrase) occurs not on the head of the proverbial pin, but in specific places – in regions, cities, towns and local commu-nities. For as Jane Jacobs (1984), the heterodox economist and urbanist, once argued, in an important sense there is no such thing as the 'national' economy, construed as a singular, coherent functioning entity, but rather a complex territo-rial system of interlocking and overlapping regional and subregional economic spaces each having its own economic structure and dynamic, each having differ-ent linkages and interdependencies with other such spaces, and all linked differ-ently into different global networks of production, trade and competition.

Of course, the regional and subregional spaces making up a 'national' economy are subject to common monetary, fiscal and regulatory arrange-ments. But not all places experience the same rate of economic growth or the same pattern of development. Economic geographers have long highlighted the systemic tendency for capitalist development to be a geographically uneven process (see Harvey, 2006; Martin, 2008: Scott, 2006). The result is the existence of spatial disparities in economic prosperity at a whole variety of geographical scales. And such disparities appear not only to be persistent, but also seemingly intractable. In the UK, for nearly a century successive governments have sought to grapple with this problem, devising various policies aimed at reducing the economic gaps between the regions. To be sure, the severity and visibility of the UK's 'regional problem' have ebbed and flowed over this time. Regional disparities were particularly acute in the economic slumps and upheavals of the 1920s and 1930s. During the long post-war 'golden age', however, regional disparities narrowed as all parts of the country enjoyed steady economic growth, so the regional problem faded to the political background. But in the 1970s, with abrupt economic slow-down, mounting inflation and the onset of deindustrialisation, regional dis-parities began to widen once more. And this widening accelerated in the

Thatcherite 1980s, fuelling a fierce debate over the existence of a north–south divide in economic performance and job opportunities (see for example, Martin, 1988, and Chapter 2 of this book). Not unexpectedly, the idea of such a divide, separating an economically buoyant and dynamic South (roughly that area south of the Wash–Severn line) from a less prosperous and economically lagging North (roughly everywhere above that line) proved highly contentious. There was dispute about whether the divide actually existed, whether it was real or imagined; about where the divide should be drawn; and about whether in reality, the map of economic disparity was more a complex 'archipelago' of local, intra-regional inequalities than a simple north–south regional divide (see Martin, 2004).

Against this background it is of more than just passing interest to examine the geographical complexion of what has been claimed by New Labour as the longest period of UK economic growth on record:

> In this my eleventh Budget, my report to the country is of rising employment and rising investment, continuing low inflation, and low interest and mortgage rates ... built on the foundation of the longest period of economic stability and sustained growth in our country's history (Gordon Brown, Chancellor of the Exchequer, *Budget Statement*, March 2007).

A key question is whether all regions and subregions of the country shared in this 'sustained growth', or whether it was primarily a feature of particular regions and cities. What in fact happened to regional economic disparities during the boom? And did the geography of the boom help shape the deep recession that brought it to an end in 2008? These are the issues this chapter seeks to address (see also Martin, 2009).

New Labour's 'long boom' and its geography

Economic growth is important because it shows whether economic activity is expanding or contracting, and it provides an indicator of the prosperity and well-being of a nation. But it is also a problematic indicator: there are empirical difficulties in measuring it accurately (and these are compounded when trying to measure it for individual regions and subregions), and in interpreting what it means for the standard of living (see Dunnell, 2009). The most common measures of economic growth are the changes in gross domestic product (GDP) or gross value added (GVA), usually adjusted for price inflation, so these are the indicators adopted in what follows.

The 'long boom' much trumpeted by Gordon Brown in fact started in early-1993, some four years prior to New Labour coming into office, as the economy began to recover from the recession of the early-1990s. It lasted up until early-2008. Whether it was indeed the longest period of sustained growth on historical record is debatable. For one thing, reliable and comparable records on GDP or GVA for the eighteenth and nineteenth centuries are not in fact available.

And in any case, the economy in that period was very different in structure and nature, so such comparisons may not be meaningful. For another, economists often refer to the 20 years or so from roughly 1950 to 1973 as a post-war 'golden age' of sustained expansion, relatively free from major recessions, so this too might be claimed as one of the 'longest' such periods on record. But leaving such debates aside, certainly in terms of growth rate achieved, New Labour can claim some credit, in that under their management the economy grew faster than under the Conservative governments of Margaret Thatcher (1979–1990) and John Major (1990–1997), and matched, or almost matched, the growth achieved in the 'golden' post-war years (see Table 3.1).

Table 3.1　Economic growth in the UK for selected periods (annual average growth of GDP at constant 2005 prices)

Period	Average annual real growth rate (%)
1950–1964	2.9
1964–1973	3.1
1973–1979	2.3
1979–1990	2.3
1990–1997	1.9
1997–2007	2.9

Source: National Statistics Online, www.statistics.gov.uk/statbase

But what of the regional dimensions of this phase of growth? There are three reasons why regional inequalities might have been expected to narrow over the course of New Labour's economic boom. First, by the late-1990s, commentators were claiming that even if the north–south divide had existed in the 1980s, it was now dead: that the intense deindustrialisation of the North during the 1980s and early-1990s had removed the source of its problem, namely an outmoded economic structure, so that its economic prospects would no longer be hampered

> The traditional 'north–south divide' unemployment problem has all but disappeared in the 1990s. This may prove to be a permanent development, since the manufacturing and production sectors, the main source of regional imbalance in the past, no longer dominate shifts in the employment structure to the same extent. Future shocks will have a more balanced regional incidence than has been the case in the past. (Jackman and Savouri, 1999: 27)

Essentially, the underlying assumption was twofold: that with their historic reliance on manufacturing dramatically reduced (see Chapter 10) and their dependence on services correspondingly enhanced, the economies of northern Britain had become much more like those of the South, so that economic growth ought to be more balanced across the regions; and, further, that services are much less prone to cyclical shocks than manufacturing, so that any downturns would be not only be less severe than previously, but also felt equally across the regions of the country. Thus, overall, regional disparities ought to narrow.

This line of reasoning was in fact closely related to a second more general argument, namely that since the early-1980s a 'new economy', driven by new information technologies, knowledge-intensive services and ideas-based and creative activities, had been developing, rapidly replacing the 'old economy' based on industry, manufacturing and mass production. Not only was this new economy supposedly revolutionising the way capitalism works – acting as the foundation for a regime of low unemployment, low inflation, high pro-ductivity growth – the claim was that it was also potentially more spatially balanced in its locational dynamics. Academics and policymakers alike believed that all regions could participate in the new 'soft economy' of knowledge-based, creative and cultural industries (Martin, 2006). The low barriers to entry, local market orientation and ease of access to universally-available information technologies characteristic of many such activities, suggested that northern regions and cities would be able to reconfigure and rebuild their economies around these new growth sectors.

And third, upon assuming office in 1997, New Labour quickly embarked on a new regional policy model aimed specifically at raising the contribution of all regions to the national growth effort, and enhancing and promoting local 'indigenous potential' was a key element of that strategy:

> The Government's central economic objective is to achieve high and stable levels of growth and employment. Improving the economic performance of every region of the UK is an essential element of that objective, firstly for reasons of equity, but also because unfulfilled economic potential in every region must be released to meet the overall challenge of increasing the UK's long-term growth rate. (HM Treasury, 2001)

The aim was 'to make sustainable improvements in the economic perform-ance of all regions by 2008 and over the long term reduce the persistent gap in growth rates between the regions' (HM Treasury, *PSA Target 2.3*). Taking these economic and policy conditions together, then, the prospects for narrowing the growth gaps between the regions might well have appeared favourable.

Actual regional experiences have not matched these expectations and aims, however. So that the geography of the recent long boom can be set in histori-cal context, Figure 3.1 shows the year-on-year cumulative differential rate of growth (that is, compared to the UK rate) of GDP for the standard regions, over the period 1980–2007. Several features are immediately apparent.

Most obviously, there have been marked regional disparities in economic growth over the past three decades. Major gaps in economic growth across the regions opened up rapidly during the Thatcher boom of 1983–1990, with London especially, but also the South-east, recording significantly higher growth than the rest of the UK, and the North-east and North–west regions

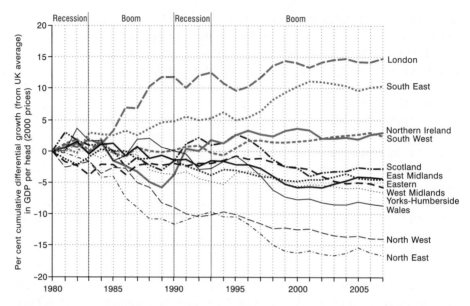

Figure 3.1 Growing apart: regional cumulative differential growth in GDP per capita, 1980–2007. Cumulative differential growth measured as cumulative annual difference between regional and national percentage change in GDP per capita. GDP measure is work-place based. (Cambridge Econometrics, http://www.camecon.com)

significantly lower (Martin, 1992). As a result of this distinct geographical bias, between 1980 and 1997 London gained more than a 22 per cent cumulative growth advantage over the North-east and North-west; and the South-east an 18 per cent advantage. Second, these differentials stabilised and even narrowed slightly during the recession of the early-1990s, when economic slowdown affected London and the South-east as much as, if not more than, it did the North-east and North-west. But, third, with the return to economic expansion after 1993, regional growth differentials widened once more. As was the case during the 1980s, the post-1993 boom was also driven mainly by London and South East, with the rest of the country – and again especially the North East and North West regions – lagging behind in economic growth. Although under New Labour the rate of divergence in growth performance across the regions did seem to slow, nevertheless by 2007 the cumulative growth advantage of London over the North-east and North-west had continued to increase, to 30 per cent, and that of the South-east to 25 per cent. Outside of London and the South-east, only the South-west region and Northern Ireland managed to record a cumulative growth outcome above the national average (the turnaround of Northern Ireland from its rapid relative deterioration during the 1980s is most noticeable). Overall, however, the data suggest that the boom that was much celebrated by Gordon Brown did little to close the long-standing growth gaps between the regions. The continued relative deterioration in economic performance of the North-east, North-west, Wales, West Midlands and Yorkshire-Humberside is especially evident.

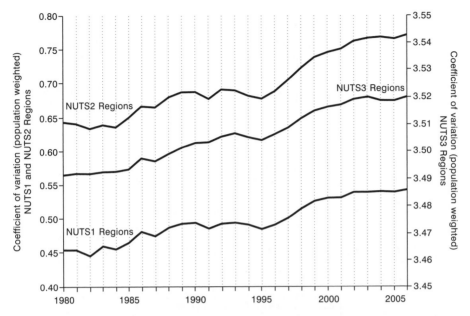

Figure 3.2 Increasing spatial inequality: the evolution of regional disparities in GDP per capita, 1980–2006. The NUTS (Nomenclature of Territorial Units for Statistics) classification of regions refers to the system used by the European Commission. NUTS1 regions are the standard, or Government Office, regions; NUTS2 and NUTS3 refer to smaller-scale subregional divisions, approximating to counties, and groups of local authority districts, respectively. (Cambridge Econometrics, http://www.camecon.com)

As a result of these trends, geographical disparities in the level of GDP per capita (as measured by the regional population weighted coefficient of variation), which had been increasing since the beginning of the 1980s, continued to widen, particularly between 1993–2002. This widening occurred at all spatial scales, not only at the level of the standard Government Office (NUTS1) regions shown in Figure 3.1, but also at subregional (NUTS2 and NUTS3) scales (see Figure 3.2). Similar divergence in GDP per capita occurred amongst the country's major cities (see Simmie et al., 2006). Contrary to what had been hoped and predicted, then, New Labour's economic boom failed to reverse the scale of regional economic disparities inherited from the 1980s. Just as in the 1980s, economic growth over the 1993–2007 period was led by London and the South-east. This geographical unevenness in economic growth raises some key issues. First, and most obviously, there is the question why economic growth was focused in these regions.

The causes of uneven regional growth

According to New Labour, the key to economic growth and success is 'competitiveness', which is defined as productivity, which in turn is viewed as

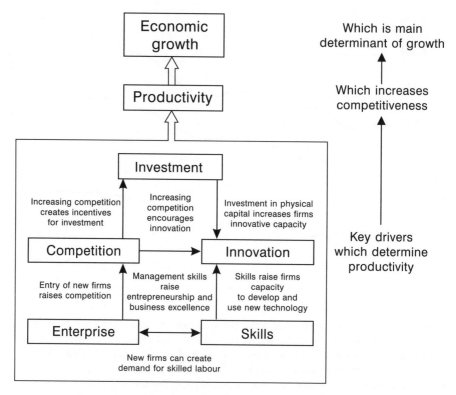

Figure 3.3 New Labour's explanation of regional growth (After HM Treasury, 2004)

being determined by five key 'drivers'. This same basic model is assumed to hold at all spatial scales, from the nation to the region, to the city, to the local authority level (see Figure 3.3). It is not altogether clear where this explanatory framework comes from, though the accent on innovation suggests a certain affinity with endogenous growth theory (back in 1994 Gordon Brown had referred to 'post-neoclassical endogenous growth theory' as a basis for government economic policy, a piece of jargon for which he was subsequently lampooned in the press). But, as it stands, this schema contains little that is explicitly spatial. It says nothing, for example, about why the supposed key drivers of growth might vary from region to region, from city to city, or why such differences may be persistent or self-reinforcing over time. Neither does it say anything about the inter-relationships and interdependencies between regions and between cities. In reality, the specific nature, strength and potential of the drivers in any region will be the complex product of the legacies of its industrial history (path dependence effects), the nature and extent of its comparative advantage in particular economic specialisms, its ability to attract capital investment and skilled and creative labour, the type of its educational, business, financial and political institutions, the scale of its own local 'home' market, and its linkages with and role within the wider global economy. And many of these factors and forces are autocatalytic, that is, mutually reinforcing (in a positive or negative direction) in their affects on a region's relative growth performance. Differential regional growth is as much about cumulative causation as it about 'indigenous potential'.

London and the South-east have for some time enjoyed a strong and competitive advantage in these terms. These regions together constitute a large 'home market' on their own account (together they make up more than one-third of the nation's population), an environment in which new firm creation thrives. They also contain the majority of the country's higher education institutions; and they attract large inflows of skilled, well-educated and creative people from elsewhere in the UK, and from overseas. As a global city, London draws in large flows of foreign capital and the offices of global companies. The South-east region – as the commuter hinterland to London – supplies a significant proportion of the capital city's workforce. And both London and the South-east have experienced industrial histories that have involved the formation of economies characterised by what we might call 'clustered diversity' – that is, numerous business clusters across a wide range of economic specialisations – which gives them a greater ability than most other regions of the country to adapt to changing economic circumstances and challenges. And of course London benefits from being the nation's seat of political and administrative power (and see Chapter 4). In short, the sort of drivers emphasised by New Labour as the determinants of regional growth have been particularly well developed and growth-enhancing in this part of the UK.

This was evident in the 1980s when the first major boom in finance and related business services occurred. And similarly, it was evident in the post–1993 'long boom', which was also rooted in finance and related activities. Nationally, output (GVA) in finance and banking increased by almost 180 per cent in real terms between 1993–2007, and by almost 150 per cent in other business services. Growth of output in manufacturing, by contrast, was a mere 11 per cent (see Figure 3.4). The expansion of finance over 1997–2007 was fuelled by a combination of globalisation, securitisation, cheap money, low inflation and 'light-touch' regulation of financial markets. Banks and other financial institutions, and the gamut of other service activities linked to these sectors, expanded everywhere across the country (and see Chapter 5). Most major cities saw their financial and business sectors expand. But it was London that dominated the process. London, of course, has long been the centre of UK banking and finance (the Bank of England was established as far back as 1694), and given its role as a leading global financial centre it generates vast foreign earnings from its banking and related activities. As a consequence, many of those working in the London banking, financial and business services economy enjoyed rapidly rising incomes, which in turn boosted local consumption and housing demand, and thus house prices. And these income and wealth effects, together with house price inflation, spilled over into London's commuter hinterland. It was not surprising, then, that whilst all regions shared in the growth of banking and finance over 1997–2007, London

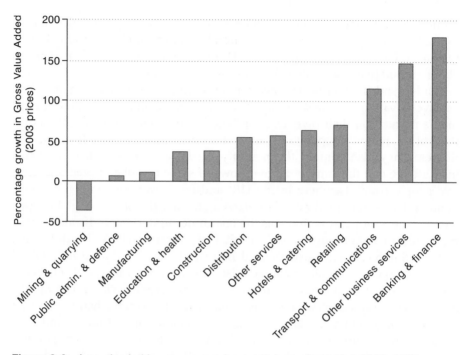

Figure 3.4 A service-led boom: economic growth by major sector, 1993–2007 (Cambridge Econometrics, http://www.camecon.com)

and the South-east in particular led the process. These two regions alone accounted for almost 50 percent of the national growth in output in banking and financial activities over that decade (see Figure 3.5). And London's share of national GVA in banking and finance increased from 29 per cent in 1997 to 35 per cent in 2007.

Likewise, London and the South-east accounted for 50 per cent of the growth of output in other business services (see Chapter 11), thereby maintaining their 40 per cent share of national GVA in this sector. The minimal growth in manufacturing over the period did little to boost the economies of those regions where this sector still remains an important, if by no means major, part of the local employment base. Indeed, northern Britain seemed to have relied much more than London or the South-east on public sector activities to underpin their growth (see Figure 3.6).

For many observers, Government included, London and the South-east are seen as constituting the 'dynamo' of the British economy, as the critical engine of national growth. Further, it is argued that this concentration of economic activity and growth is beneficial for the rest of the economy: not only does growth diffuse out to the other regions, London and the South-east act as an important market for many goods and services provided by the other regions. Some estimates suggest that London alone each year imports over £120 billion of goods and services from the rest of the country, and that it

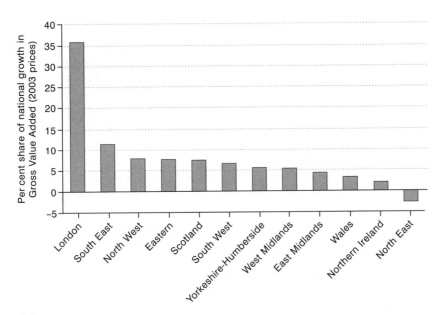

Figure 3.5 London's dominance in the financial services boom: shares of national absolute growth in GVA in finance and banking, 1997–2007 (Cambridge Econometrics, http://www.camecon.com)

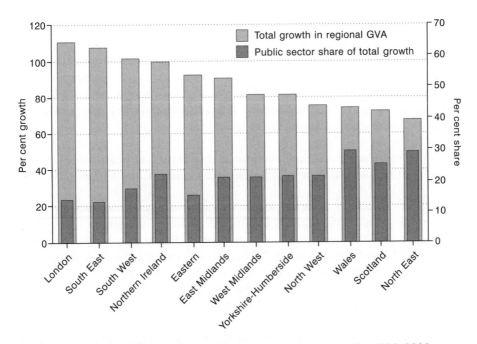

Figure 3.6 The contribution of the public sector to regional growth, 1999–2006 (National Statistics Online, www.statistics.gov.uk/statbase)

makes a net annual contribution of around £13 billion to UK public finances (City of London, 2007).

This creates a major policy dilemma (Martin, 1993). On the one hand, it could be argued that national and regional policy should do nothing to counter or stem the growth dynamic of London and the South-east, since this would reduce national growth as a whole. In other words, the agglomeration of economic activity in London and the South-east results in national efficiency gains. To this end, New Labour was concerned that any supply-side constraints – such as housing shortages – should not impede the economic growth in this part of Britain. For this reason, it imposed a massive house-building programme in the region over the next couple of decades. But, on the other hand, fostering the continued agglomeration of economic activity, and of skilled and educated workers, in the London–South-east area may do little to help the rest of the country to catch up in terms of growth rates and wealth creation. The hope may have been that this outcome can be avoided by a regional policy aimed at enhancing the 'indigenous' growth potential of the regions, and by promoting major agglomerations of economic activity in northern Britain (such as the 'Northern Way'). But a view has nevertheless emerged in certain quarters that there is some sort of trade-off between national efficiency and regional equality, and that the continued agglomeration of activity in London and the South-east is essential for the former. As a recent Treasury paper puts it:

> Theory and empirical evidence suggests that allowing regional concentration of economic activity will increase national growth. As long as economies of scale, knowledge spillovers and a local pool of skilled labour result in productivity gains that outweigh congestion costs, the economy will benefit from agglomeration, in efficiency and growth terms at least ... policies that aim to spread growth amongst regions are running counter to the natural growth process and are difficult to justify on efficiency grounds, unless significant congestion costs exist. (Lees, 2006: 24)

And an even more extreme version of this view has been voiced by the centre-right think tank Policy Exchange, in its report *Cities Unlimited:*

> There is no realistic prospect that our [northern] regeneration towns can converge with London and the South East. There is, however, a very real prospect of encouraging significant numbers of people to move from those towns to London and the South East ... The implications of economic geography for the south and particularly South East are clear. Britain will be unambiguously richer if we allow more people to live in London and its hinterland. (Leunig and Swaffield, 2008: 5–6)

Such arguments can be contested not just on social equity grounds – do we really just let northern cities and communities decline and atrophy? – but also on sustainability grounds: is continued concentration of growth, economic activity and population in London and the South-east really sensible, given the impacts of such agglomeration on costs, the environment and the quality of life in this part of Britain? And in any case, excessive spatial agglomeration can be a source of national economic *in*efficiency and instability, as the end of the long boom has demonstrated.

From boom to bust: the regions and the recession

Just as Gordon Brown as Chancellor repeatedly claimed that the 'longest boom' in our history was a domestically grown triumph, in large part due to his policies of economic stability and financial prudence, so as Prime Minister he repeatedly attributed the recession that began in 2008 to external 'global' forces beyond the UK's control. It is true that the global 'credit crunch' which in its turn precipitated global recession began life in the collapse of the 'sub-prime' mortgage finance market in the US. But the collapse of the banking sector and the housing market in the UK, and the recession that followed, cannot simply be attributed to 'external forces'. British banks and mortgage lenders were not innocent bystanders or victims of the drama that unfolded over 2007–9. They were active participants. Encouraged by the UK's lax national financial regulatory regime, by low interest rates, and by what appeared to be low inflation (on account that the official measure excluded housing costs), the banks and demutualised building societies went on a mortgage lending spree, part of which was financed by borrowing in global whole-sale money markets (via complex financial instruments such as mortgage-backed securities). Between 1997 and 2007, a record £1.2 trillion of new mortgage loans were made. Some £0.522 billion of this total was accounted for by London and the South-east. This huge mortgage bubble, and the rapid increases in house prices, enabled households to extract record sums of equity from their housing assets, money that along with rising real incomes contributed to the boom in consumer spending. Again, London and the South-east led this process.

These interacting developments hugely benefited the financial institutions and financial classes in London and its environs: as we have seen, finance and banking were the main sectors that led the long boom. In fact, the house price and mortgage bubble originated in London and the South-east. House prices in these regions took off in 1995–96, and especially after 1997–98, but it took 3–4 years for the bubble to diffuse out to northern regions of the country. That the house price bubble started in London and the South-east was unquestionably stimulated by the faster wage and earnings growth in these areas, helped by London's position in what in many ways has become a global labour market for highly-paid financial and professional business service workers. The generous annual bonus culture that came to be an integral part of London's financial worker class also played no small part. As the Nobel economist James Tobin (1984) argued some 25 years ago, having a global financial centre at the heart of a leading growth region can be a source of national economic inefficiency, since it tends to exert additional upward pressure on local wages, salaries and costs which then spreads out across the country as a whole. Further, while such a centre – like London – may generate valuable earnings from financial trading, it also acts as a portal through which instabilities and shocks originating in the global financial system are readily imported and transmitted across the economy.

When the 'subprime' bubble burst in the US it quickly crossed the Atlantic to throw those UK banks and demutualised building societies which had raised funds by selling mortgage-backed securities on the global market into crisis. As the crisis spread throughout the UK financial system, and across much of the global system, the onset of economic recession quickly followed. The long boom had turned into bust.

In speed and scale, the current contraction is certainly one of the worst of the post-war period. How long it will last is less certain. At the time of writing, the recovery has yet to gather momentum, the state of the economy remains extremely fragile and future prospects uncertain. Since the recession has been driven in large part by the credit crisis and the collapse of the banks, the expectation was that its impact would likewise be most strongly felt in the finance sector. This in turn implied that London and its hinterland would also be hardest hit. In November 2008 the Mayor of London predicted that some 70,000 jobs would be lost in London's financial district over the coming year because of the crisis. It would be surprising if a major shakeout of financial and banking labour in London did not occur. But as the banks hit by the crisis have restructured and in some cases merged, the resultant job losses in the finance sector have affected other regions as well as London. Further, given the rapid slowdown in domestic demand, and the near-drying up of overseas export markets due to the global nature of the recession, British manufacturing has also been badly impacted. The available evidence suggests in fact that the slowdown in growth has been much more pronounced in manufacturing, which was in a much less strong position to begin with (see Figure 3.7).

Given that reliable official data on key indicators of regional economic performance, other than unemployment, are only published after a considerable time lag, and that at the time of writing the economy is barely out of recession, mapping the impact of the current downturn is not easy. Table 3.2 gives some very rough estimates of the decline in GDP by region over 2008 (1st quarter) to 2009 (1st quarter). These should be treated with considerable caution (for the reasons set out in Table 3.2), but they do suggest that while the downturn has been sharp and pronounced everywhere, thus far London and the South-east seem to have been affected no worse than other regions. The picture with regard to unemployment is more mixed, though here too, to date London and the South-east do not appear to have witnessed any more severe decline than

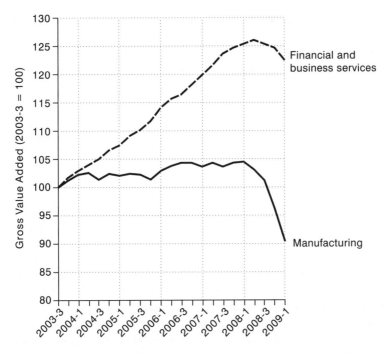

Figure 3.7   The onset of recession: output in finance and business services versus manufacturing
Source: Gavurin (http://www.Gavurin.com) and ONS (National Statistics online, http://www.statistics. gov.uk/statbase), 2009

Table 3.2   The impact of the recession on the regions: estimated decline in output, and the change in unemployment rate, 2008 (1st qrtr) to 2009 (1st qrtr)

Region	% decline in GDP	Absolute increase in unemployment rate
London	−4.41	1.8
South–east	−5.31	1.4
Scotland	−5.76	1.3
Eastern	−5.78	1.4
South-west	−5.90	2.1
North-west	−6.07	1.9
North-east	−6.17	2.2
Yorkshire-Humberside	−6.26	2.9
West Midlands	−6.28	3.0
East Midlands	−6.47	1.6
Wales	−6.50	2.3
Northern Ireland	N/A	2.2

Note: The changes in GDP by region are derived by applying national estimates of the change in output by major sector to the sectoral employment structures (as at 2007) of the regions. They thus assume that each sector has responded to the recession similarly in every region, and that regional employment structures have remained unchanged over the recession period. For these reasons, in addition to the fact that the national estimates are themselves preliminary, these figures should be regarded as highly provisional.

elsewhere; whereas the West Midlands and Yorkshire-Humberside seem to have been hardest hit.

It is too early to judge what precise impact the current recession and its aftermath will have on the future evolution of regional economic disparities. One of the stated aims of New Labour was

> to make sustainable improvements in the economic performance of all English regions by 2008, and over the long term reduce the persistent gap in growth rates between the regions, demonstrating progress by 2006. (DTI *PSA Target 7*; HM Treasury *PSA Target 2.3*; DCLG *Target 2*)

Just how long the 'long term' was remains unspecified, but it is difficult to argue that much progress had been made by 2006 in reducing regional disparities. There was talk of signs of a 'turn around' in the North-east's relative growth trend after 2003, yet as Figure 3.1 shows, this looks to have been a temporary phenomenon, just as the more marked relative improvement during the recession of 1990–1993 subsequently proved to be. The fact is that London and the South-east appear to be much more resilient and robust economies: these regions have led the growth process in the last two booms, and seem better able to recover from recessions than other regions. Much will depend on how the financial and banking sectors respond. Further, given the scale of Government borrowing that has been needed to bail out the banks and stimulate the economy ('quantitative easing'), major cuts in public expenditure will be required to reduce the historically large deficit, and this could impact differentially across the regions, and particularly those that have been more dependent on the public sector for their growth over the past decade or so (see Figure 3.6). Unless the next recovery, when it comes, is based on a vastly different type of economic dynamic, the cumulative growth gaps between the regions are likely to persist for some time to come.

A concluding comment

The experience of the UK economy over the past decade and a half provides a striking illustration of how phases of boom and bust in the national economy are in fact geographically uneven processes. Regional disparities in growth and prosperity have long been a systemic feature of the British economy, despite nearly a century of policies intended to reduce them. Since the beginning of the 1980s these disparities have widened, and the celebrated long boom from 1993–2008, a period dominated by New Labour in government, saw no marked closing of the growth gaps between the regions. The fact is that there are very strong forces making for the persistence of regional differences in economic growth. Explaining the nature of those forces is one of the challenges that economic geographers continue to grapple with.

Further reading

- Martin (2004) provides a wide-ranging discussion of the debates surrounding the North–South divide in economic and social prosperity in the UK, and explores both the myths and realities surrounding the divide.
- Martin (2006) examines how a new regime of growth in the UK economy has reinforced rather than reduced economic, social inequalities and regional inequalities.
- Leunig and Swaffield (2008) set out the highly contentious argument that because previous attempts to regenerate the economically depressed North of Britain have had only limited success, and because such policies in any case work against market forces, policy should instead encourage and facilitate the continued market-driven concentration of economic activity and growth in Britain's South, and let many northern cities decline. The reader hopefully will see the flaws and injustices of such a proposal.

References

Brown, G. (2007) *Chancellor of the Exchequer's Budget Statement*. London: HM Government.

City of London (2007) www.cityoflondon.gov.uk/Corporation/media_centre files2007/A + capital + contribution + + Londons + Place + in + the + UK + Economy + 200 7-08.htm (accessed June 2008).

Dunnell, K. (2009) Measuring economic performance, *Economic and Labour Market Review*, 3: 18–30.

Harvey, D. (2006) *Spaces of Global Capitalism: Towards a Theory of Uneven Geographical Development*. London: Verso.

HM Treasury/DTI (2001) *Productivity in the UK: 3 – The Regional Dimension*, London: HM Treasury, p. v.

HM Treasury/DTI/ODPM (2003) *A Modern Regional Policy for the United Kingdom*. London: HMSO, p. 14.

HM Treasury (2004) *Developing Decision Making: Meeting the Regional Economic Challenge – Increasing Regional and Local Flexibility*. London: HM Treasury.

Jackman, R. and Savouri, S. (1999) Has Britain solved the 'regional problem'? in P. Gregg and J. Wadsworth (eds), *The State of Working Britain*. Manchester: Manchester University Press.

Jacobs, J. (1984) *Cities and the Wealth of Nations: Principles of Economic Life*. New York: Random House.

Lees, C. (2006) Regional Disparities and Growth in Europe, *mimeo*. London: HM Treasury.

Leunig, T. and Swaffield, J. (2008) *Cities Unlimited: Making Urban Regeneration Work*. London: Policy Exchange.

Martin, R.L. (1988) The political economy of Britain's North–South divide, *Transactions of the Institute of British Geographers*, NS 13: 389–418.

Martin, R.L. (1992) The economy: has the British economy been transformed? Critical reflections on the policies of the Thatcher era, in P. Cloke (ed.), *Policy and Change in Thatcher's Britain*. Oxford: Pergamon. pp. 123–58.

Martin, R.L. (1993) Reviving the economic case for regional policy, in R.T. Harrison and M. Hart (eds), *Spatial Policy in a Divided Nation*. London: Jessica Kingsley. pp. 271–90.

Martin, R.L. (2004) The contemporary debate over the North–South divide: images and realities of regional inequality in late-twentieth century Britain, in A.R.H. Baker and M.D. Billinge (eds), *Geographies of England: The North–South Divide, Imagined and Material*. Cambridge: Cambridge University Press. pp. 15–43.

Martin, R.L. (2006) Making sense of the new economy? Realities, myths and geographies, in P. Daniels, A. Leyshon, M. Bradshaw and J. Beaverstock (eds), *Geographies of the New Economy: Critical Reflections*. London: Routledge. pp. 15–48.

Martin, R. L. (2008) Geographies of Economies, in R.L. Martin, (ed.) *Economy: Critical Essays in Human Geography (Vol 8 in Contemporary Foundations of Space and Place Series)*. Aldershot: Ashgate, pp.11–37.

Martin, R.L. (2009) The recent evolution of regional disparities – A tale of boom and bust, in J. Tomaney (ed), *The Future of Regional Policy*. London: Smith Institute, p.13–24.

Scott, A.J. (2006) *Geography and Economy*. Oxford: Oxford University Press.

Simmie, J., Martin, R.L., Wood, P., Carpenter, J. and Chadwick, A. (2006) *The Competitive Economic Performance of English Cities*. London: Department of Communities and Local Government.

Tobin, J. (1984) On the efficiency of the financial system, *Lloyds Bank Review*, June.

Acknowledgements

This chapter is based on a much longer presentation originally made at the session on 'The Economic Geography of the UK' at the Annual Conference of the Royal Geographical Society-Institute of British Geographers, August 2008.

PART 2

LANDSCAPES OF POWER, INEQUALITY AND FINANCE

4

THE CITY AND FINANCE: CHANGING LANDSCAPES OF POWER

John Allen

AIMS

- To introduce and describe the dominant role of the City of London in the context of the UK economy

- To explore how the City of London has reproduced this dominance and whether the 2007–09 economic downturn signals its decline

- To examine the basis of the City of London's power and its contested nature

At the end of the first decade of the twenty-first century, the landscape of financial and economic power across the UK economy looked markedly different from that at the turn of the century. Back then, the business of banking, broking and insuring, fuelled by deregulation and the globalisation of capital flows, accounted for a growing proportion of the UK's wealth and profitability. The sense in which the UK economy could turn its back on its manufacturing past and embrace a more weightless, financially driven future spearheaded by the City of London's financial institutions looked not only convincing – among policymakers, at least – but also such a belief had assumed the status of an economic creed. The City, a shorthand for the UK's financial sector (although see Chapter 5), represented the money-making future that was to cement the UK's role in the global economy. Fast-forward to the end of the noughties, such beliefs not only looked far from convincing, they also drew into question the very basis upon which the UK economy rests.

Amidst the battered landscape of London's financial sector, the once mighty banks, investment houses and trading firms struggled to come to terms with the financial storm that had wrecked their businesses. The sub-prime crisis in the US, the trigger for the enveloping storm, inflamed by a culture of excessive risk-taking, had altered the financial landscape almost beyond recognition. Jobs in the City, from a peak of 360,000 in 2008, looked set to shrink by one third, cumulative losses and write-downs for the world's largest financial institutions predicted to reach $1000bn and the contribution of London's financial services to the nation's wealth forecast to be slashed from 9 to 6 per cent of GDP. The dominance of the City looked to be over: its power greatly diminished, its authority and influence no longer recognised. Or so it seemed to some at the time.

Opinion is divided on the matter, largely over how best to interpret the 'highs' and 'lows' of power when the economy changes abruptly and business shrinks in both scale and volume. What significance should we attach to the changing momentum of the UK economy for, say, the big banks as a dominant force in economic markets and beyond? In such circumstances, what happens to their power and influence over the government's longer-term economic strategy? When, for example, the US bank Lehman Brothers slid into bankruptcy in September 2008 after 158 years of trading, did its power simply evaporate? And if so, what about the other big overseas investment houses still trading in London's Canary Wharf? Has their authority simply taken a knock, only to be restored in an economic upturn when their 'reserves' of power can be drawn upon once again? Or has the City of London's power and influence diminished irrevocably in the financial crisis of the late 2000s?

As ever, it depends. It depends upon how financial and economic power is understood; how the way that power 'works' is read. In this chapter we look at two contrasting interpretations of power and finance, each resting upon a particular view about how the economy changes and how it is organised. The first, broadly *structural* explanation of events, focuses upon the financial system as a whole, where the power of City institutions stems from their distinct position in the circuits of credit, finance and capital. In direct contrast, the second account rests upon a more *networked* account of events which draws attention to the largely expedient actions of financial elites in skewing rewards of power and wealth to their advantage. Where one would consider the latest financial crisis as a periodic punctuation in the City's dominance, the other would see the evident loss of the City's authority as the result of over-reach and misdirection with no guarantees as to the future abilities of its financial institutions.

Before we look at each 'reading' of power and finance in turn, we consider the broad historical shifts in the City's ability to reproduce itself as a powerful force, in relation to both economic markets and the state.

From dominant to bankrupt force?

Whilst there is disagreement over *how* the City has reproduced its dominance over time, there is broad agreement over the fact of that dominance – for the time being at least. What is meant by power in this context is the ability of the City to bend or influence the will of others to gain advantage. The nature of that advantage is implicitly financial: to profit at the expense of industry and shareholders, or consumers and taxpayers, and the like, by forcing a redistribution in the City's favour. In short, it comes down to the ability of the City to skew the system to its advantage by ensuring that others have no choice but to fall into line. Notwithstanding the fact that the City has altered in both its economic character and size over the past two centuries, the extent to which it has maintained its dominance over time is often attributed to its core commercial activities: the ability of its institutions to place themselves in the middle of all kinds of complex financial activities and to benefit disproportionately from that role.

With respect to this commercial dominance, there has been a long-standing debate about the 'exceptional' character of the UK's economy which resurfaced in the late 1980s with an even stronger emphasis upon the pre-industrial stamp of the nation's ruling block and the hegemonic role of the City. Also at issue in this debate, however, was the precise role that the City performed in relation to the 'real' economy of the UK, in particular its disappearing manufacturing base. In opposition to those who portrayed the City as a type of finance capital, orientated to global markets and largely separate from the domestic parts of the UK economy, Geoffrey Ingham (1984, 1989) argued pointedly that the City's earnings came predominantly from its commercial role.

By this, he meant that the City's profits from the 1830s onwards stemmed overwhelmingly from the ability of its institutions to act as 'middlemen' or brokers within the financial system. By inserting themselves in the middle of the foreign exchange and money markets, commercial brokers profited from their role as intermediaries on a fee or commission basis. As traders of capital, stocks, bonds and securities, City firms built up an historical advantage in the exchange and distribution of economic resources that witnessed their ever growing significance to the UK economy. The economic historian C.H. Lee, in his study *The British Economy Since 1700* (1986), charted the rise of this commercial significance, dating back to the mid-nineteenth century, which showed a southern service economy, centred on London and commerce, consistently generating around half the national rate of economic growth and the bulk of the employment growth (see Figure 4.1).

Economic significance, however, is not the same as economic dominance, and although work by other economic historians such as Cain and Hopkins' (1987) account of 'gentlemanly capitalism' point towards the dynamic role of the City of London in the national economy from the 1850s through to the Second World War, the argument does not amount to a justification of the City's *persistent* dominance. The ability of the City to skew the economic system to its advantage, to shape the conditions of existence for everyone else, as a continuous feature, sits uneasily alongside just some of the historical, global events of the last century which damaged its fortunes: the devastating impact of the First and Second World Wars, for example, or the demise of the gold standard and the 1930s global depression.

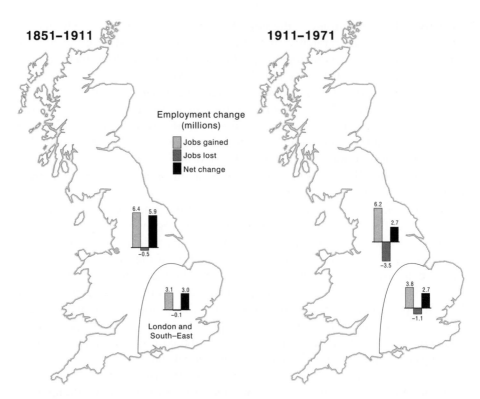

Figure 4.1   London and the south-east's regional pre-eminence – employment change 1851–1971 (adapted from Lee, 1986: 264–5)

Ingham acknowledges the significance of these events to the shifting fortunes of the City over time, yet he still wishes to maintain that the commercial dominance of the City remained throughout the twentieth century despite the obvious economic difficulties it faced, including the growth of opposition from industry and manufacturing. His claim is not that the UK had grown a top-heavy financial system at the expense of industry, but rather that the network of institutional relations between the City and government, in essence the relationships between finance houses, the Treasury and the Central Bank, worked in favour of the City's commercial interests over time. Ingham is at pains to point out, however, that this state of affairs was not down to any capability on the part of the City to wield power 'over' the state, but that the different interests of all those involved coalesced around a relatively stable monetary system. While, from the City's standpoint, instability can lead to speculation and short-term gains, in the long run the stability of monetary system engenders trust and confidence in the commercial markets.

Another way of saying this is that despite the often divergent agendas of government and financial bodies, effectively they all enjoyed the benefits of a positive-sum game. Yet if that interpretation is correct, to what extent can the City be said to have forced a redistribution in its favour by bending the will of others? If the City has been able to reproduce its dominance throughout the twentieth century, and indeed before that, then its role as commercial broker and clearing house should have had some bearing upon its ability to skew rewards in its favour. From the late 1950s and early 1960s on, with the shift in the character of the City towards a more supranational orientation, that ability became markedly evident.

In the latter part of the twentieth century, a number of developments served to reproduce the City's privileged position in the UK economy, the most significant of which spoke to the increasing internationalisation of the City's activities and the rise of a certain kind of commercial expertise and professionalism quite different from earlier forms of 'gentlemanly' conduct (Thrift, 1994). The internationalisation of bank lending, via the Euromarkets, together with the development of the euro currency markets, witnessed a rapid increase in the number of foreign banks operating in the City. Alongside the internationalisation of London's equity markets brought about by the big bang reforms of 1986, and the relaxation of barriers to trade within the financial markets, the liberalisation and deregulation of the City's activities saw an unprecedented explosion in the flow of funds and capital through the City. Changes of this magnitude and scale were reflected in the ballooning global transaction figures where those mediating these voluminous financial transactions found themselves enjoying positions of considerable power and advantage.

The power in question arose from the role performed by commercial brokers and financial traders, but to refer to that role as one of straightforward domination and control is perhaps to miss the point. In many respects the advantages enjoyed by financial intermediaries are not particularly new; it concerns the unequal access to the quantity and quality of information available to them, in contrast to industrial corporations and pension funds, as well as other investors such as mutual funds, university endowments and public sector agencies. With the routine allocation of capital and the management of risk, such basic asymmetries are not necessarily an issue, but in the increasingly volatile international financial environment of the 1970s and 1980s the diversification of risk, rather than its elimination, introduced an element of uncertainty and financial innovation in the system that favoured those best able to engineer financial solutions (Pryke and Allen, 2000).

One of the effects of this environment was to produce 'new' and restructured financial instruments that were suitably tailored to the uncertainties of the monetary system taking shape. The use of such instruments, the ability to bundle up products to distribute risk and leverage future income streams, has been a key driver of the growth of 'financialisation' over the past two decades. Financialisation, according to Robin Blackburn (2006: 39), can be defined as 'the growing and systemic power of finance and financial engineering' and although not a novel phenomenon it differs from previous rounds of financial innovation in the sheer scale and expansion of its engineered 'products'. Almost anything it seems can be turned into a bond-like security – mortgages, student loans, credit card debt, health insurance, toll road fees, utility bills and the like – sold on and used as collateral to generate further speculative activity and risk abatement (Leyshon and Thrift, 2007).

The growth of 'secondary trading' and the ability to engineer ever more elaborate instruments to repackage loans led to significant economic gains for those dealing with the risks, but a disproportionate share of those gains, as Blackburn points out, has been swallowed by those actively managing and trading the risks. Those in the middle, the traders and brokers, exploited their informational advantage to skew financial rewards in their favour. Financialisation, it seems, is but the latest move by which the City has been able to reproduce, if not its straightforward dominance, then its ability to engineer outcomes which disproportionately benefit its institutions. That is until recently, when the actions of those very same institutions brought the financial system to the edge of collapse and bankruptcy (Blackburn, 2008).

As in previous eras, it is difficult to suggest that the City, in getting its way on financial remuneration, exerted power over the institutions of the state. Recent governments have supported the view that the UK had a comparative advantage in financial services and that a competitive, risk-taking City was a central component of that economic policy (see Table 4.1). The relentless expansion and profitability of the UK's financial sector from the 1960s and 1970s on, reinforced by the deregulation of the City, was not out of line with the Central Bank's inflationary concerns, nor at the expense of the Treasury's interests. But that expansive growth was also accompanied by an institutional momentum and economic volatility that ushered in a culture of risk-taking which, in hindsight, favoured the City at the expense of just about every other stakeholder – shareholder and taxpayers alike. That that heady momentum came to an abrupt halt in 2008, and with it the bewildering array of risk-taking gains, has left the City in a position where its dominance is once again in question. If its moment of financial hubris is over, does the City's nemesis spell out its inevitable loss of leverage and authority? Or are we simply witnessing the latest bout of market readjustment after the bursting of a financial bubble?

Table 4.1 Selected financial indicators – top 10 cities, 2008

	Total value of equities trading	Total no. of derivatives contracts	Total no. of commodities contracts	Banking/financial services companies	Investments/ securities firms	Total value of bond trading
1	New York	Seoul	New York	London	New York	London
2	London	Chicago	London	New York	London	Copenhagen
3	Tokyo	Frankfurt	Chicago	Tokyo	Tokyo	Madrid
4	Frankfurt	London	Shanghai	Hong Kong	Hong Kong	Moscow
5	Shanghai	Philadelphia	Tokyo	Frankfurt	Singapore	Bogota
6	Singapore	Mumbai	Mumbai	Singapore	Chicago	Istanbul
7	Paris	Sao Paulo	Osaka	Paris	Paris	Seoul
8	Milan	Johannesburg	Kuala Lumpur	Shanghai	Seoul	Frankfurt
9	Hong Kong	New York	Sao Paulo	Milan	Frankfurt	Milan
10	Shenzen	Mexico City	Johannesburg	Madrid	Madrid	Tel Aviv

Source: Adapted from Sassen, 2009

Put another way, is this the upending of the City's established dominance, when a broken industry comes to terms with its demise as a powerful economic force? Or is its dominance merely on hold, whilst its banks sit out the cycle of blame and await the next upturn in its business and economic fortune? How you answer these questions turns upon how you read the landscape of economic and financial power.

The City as a systemic bloc of power...

One reading, which reaches back to various Marxist accounts of the basis of power in capitalist economies like the UK, focuses upon the ability of the City to reproduce its power through a pattern of *structural* dominance. By structural in this context is meant that the City, as an institutional bloc of power, is able to place limits upon what is and what is not possible in the economy as a whole. Whilst there may be periodic shifts in the ability of bankers and other financial actors to shape the surrounding business environment, the existing structures of the UK's capitalist economy reproduce the power of the City over time. The enduring, yet changing pattern of the City's dominance, on this view, operates as a *systemic* feature, independent of anybody's will.

This brings to the fore the notion of financialised capitalism. Earlier I mentioned the ability of the City to skew the system to its advantage by ensuring that others have no choice but to fall into line. The ability to constrain the free choice of others by leaving them no room for manoeuvre is the defining characteristic of *domination.* When that dominance is structural it implies that whilst government bodies and regulating authorities may curtail, say, the most recent excessive risk practices of the City, it can only do so within an economic framework that leaves the financial system as a whole broadly intact. With finance in private hands, bankers and financiers operate the economy's money and that very fact places them in a structural position quite different from manufacturing or retail firms, for example. The power of the City in this sense, according to Lawrence Harris (1988), stems from its distinct position in the international circuits of capital; first, through its control over the use of money as a means of circulation, and second, through its control over highly mobile, supranational financial assets.

In relation to the first point, we are close to Ingham's stress upon the commercial role performed by the City, whereby the distinct position occupied by financial traders and brokers places them in the special role of buying, selling and transferring the economy's money. Effectively, their power is exercised by dealing in money, credit and financial instruments, not merely as a monopoly provider of such services, but as the point through which such practices of money-dealing have to pass. They are able to skew rewards in their favour through their ability to appropriate profits from the spread of the buying and selling prices of foreign exchange, equities and bonds, as well as from the fees and commissions charged for the transfer, exchange and management of money balances.

From one viewpoint, this simply amounts to a bundle of lucrative trading practices. But when considered from the vantage point of the financial system as a whole, rather than, say, through the eyes of dealers and investment managers at the US giant Citigroup, or the Swiss Bank UBS, this advantageous structural position is precisely what has enabled the City to reproduce its dominance over time. Foreign banks in the City are not powerful simply because of their international reach; they are powerful because of the position they occupy in the circuits of capital.

This is not to deny that individual investment managers are able to exercise power at the expense of savers or trust funds, for example, but that they do so as part of a financialised form of capitalism which has benefitted the City disproportionately. Over time, on this view, the development of London as an international financial centre, through its accumulated expertise and innovative edge, has enabled it to be at the forefront of the recent wave of financialisation which has spread risk across the globe. This supranational growth of the financial system, with the City as one of its chief engineers driving liquidity and mobility, witnessed the gains from dealing with risk rise exponentially to a level where those positioned to skim the rewards enjoyed extravagant remuneration. That is, until the momentum in the system stopped in the late 2000s.

When the power of the City is thought of as systemic, the implication is that the structures of financialised capitalism reproduce its dominance despite fluctuations in the fortunes of its banks and finance houses. There is, in effect, a systemic bias in favour of the City as a whole. This bias, crucially, is not one that can simply be challenged and overruled, but is part and parcel of the way that the UK's capitalist economy operates; the way that its economic relations are structured and organised. If this sounds a little rigid, the intention is not to suggest that government authorities, for instance, are powerless to counter the dominance of the City; rather it is to point out that economic growth and profitability within this framework is *constrained* by the powers of finance.

Money is the bedrock of the whole economic system and those in a position to control it – the banks, the investment houses, and especially those who allocate capital and manage risk – operate it largely to their advantage. Moreover, they do so as part of a structural configuration which situates them, as well as governments, industrial corporations, pension funds and other institutional actors, within a pre-existing set of economic relations that constrains and sets limits upon what can change. Even if the whole banking system becomes more regulated and less lucrative as various financial bubbles burst, the underlying imperative of financialised capitalism – that finance becomes ever-more liquid and independent from production – enables the City to enjoy a dominant economic position that others are forced to confront.

On this interpretation, the economic system itself confers power and resources upon some actors, but not others, and likewise is capable of forcing a redistribution in favour of some interests at the expense of others. The latest go-getting, deal-making culture of excessive risk may be discredited for now, but the structural imbalance of power between finance and the rest of the economy remains in place.

...or a more tenuous, networked power?

Critics of a systemic view of the City's power point to its over-blown character, where it would seem that little can ever dent the ability of its financial institutions to grasp the lion's share of economic gains. This overly deterministic view is countered by one that stresses the *tenuous* nature of the City's powers; that whatever dominance its banks have been able to enjoy represents an achievement, not a given. Dominance, on this view, is something that is made and remade; a precarious achievement subject to the networked abilities of a multitude of bankers, brokers, dealers and investment managers which hinges upon their own sense about how it is put together. If, until recently, the City has been seen as a financial powerhouse, that is not because of any enduring quality, but rather because its financial elites have, over time, been more or less successful in engineering outcomes to suit their interests. The leverage of such elites is largely expedient: it stems from a practised ability to manipulate financial transactions to their advantage or risk losing that power and advantage in the process.

If the power of the City's traders is more tenuous, less substantial, than often perceived, that is because it is what they *do* that matters, rather than any reserves of power that may be called upon because of their economic position. Embedded position is still significant to those who draw attention to the *exercise* of power and authority, but primarily because it points to the networked role performed by brokers, bankers and the like. With the onset of financialisation, such individuals found themselves well placed as intermediaries to exploit resource asymmetries to their advantage. But such potentially rewarding roles, as noted earlier, are not new to the City. What is new, each time, is their achievement; that different financial elites have been able to skilfully exploit whatever advantages arise to enrich themselves at the expense of others.

Understood in this way, the power of the City from the nineteenth century onwards is not one thing; it comprises a variety of financial elites, both old and new, who have been able to use their trading roles to skew rewards in their favour. At times, this has been achieved through acts of domination, where investors have no choice but to pay exorbitant fees and commissions; at other times, the engineering of financial deals and the temptation of rewards too great not to want speaks more to an agenda of manipulation and inducement. Different registers of power and influence, in that sense, have come into play at different historical moments, and are enacted by diverse financial elites. Cain and Hopkins' gentlemanly elites shared a common educational and class background based upon personal ties, whereas today's professional elites, according to Savage and Williams (2008), operate in a looser, networked fashion, often connected through a third party than any direct tie.

For Savage and Williams, the direct personal ties which characterised the old established City elites are of secondary importance in today's global networks, where the ability to broker or bridge the holes in monetary networks holds the key to a profitable performance. Separate professional intermediaries working at, say, the German bank, Commerzbank or the French bank, Societé Générale, may 'work' the net in similar ways, putting together innovative financial deals around the globe that, for example, connect previously unrelated investors through the secondary trading market. The ability of such professional elites to enrol institutional investors, as well as wealthy individuals, to translate and align their different interests, and to 'fix' an overall orientation is, on this account, a form of networked power (see Allen, forthcoming).

Such mediating elites do not compel others to enrol in such networked ventures. In recent years, the win–win situation that securitised loans and the sophisticated management of risk presented to players was, in itself, a sufficient inducement. The fact that such gains turned out to be illusory for some, if not most of the investors, merely points to the contingent nature of the City's power. When the networks start to fall apart, so too does the power and influence of those who previously held them together. The loss of authority experienced by investment bankers and financial analysts after the recent meltdown in the City's fortunes is, in that sense, following Bourdieu (1989), a loss of symbolic power, where trust and social recognition simply disappear.

Of course, such groups continue to occupy a distinct position in the circuits of capital, but there is no guarantee that their future actions will do anything other than confirm that loss of power. Money may be the bedrock of the economy, but if risk is mismanaged and capital misallocated on a grand scale, as happened in the 2000s, the gains and influence achieved in an earlier period simply fade away. In that respect, the expedient actions of financial elites, in hindsight, turned out not to be quite so conducive to the needs and demands of the moment.

The implication of this is that power of the City, on this view, turns out to be a tenuous achievement; it is not something held by financial institutions, rather it is produced through networked forms of interaction. Such powers of finance may be generated by the application of resources and skills over tracts of space and time, and expand or contract in line with the monetary resources available *and* how they are used. There is a pragmatic side to things, in so far as financial elites may misuse or simply waste resources, or fail to recognise that what works well in one context may fail in another, and in so doing find that their power and influence shrinks in both scope and scale (see Allen, 2008).

On this interpretation, the ability of bankers and brokers to over-reach themselves in a moment of financial volatility and excessive momentum may or may not upend the established City order, but it does spell out an evident loss of authority – for now at least. As for how long, that is a contingent question, as there is no guarantee that in any future economic upturn City bankers will be able to enrich themselves to such an extraordinary degree as over the past two decades. The provisional and contextual nature of their power means that it is simply not possible to know in advance whether the City of London's power and influence diminished irrevocably in the financial crisis of the late 2000s.

The issue is further complicated by the fact that, for much of the time, what actually passes for power in the City is not the kind that bends the wills of others, to gain advantage. In no small part this is because the power exercised by financial elites is largely directed at working the networks: forging connections, bridging the gaps, and stabilizing interests so that associations hold together. As such, it is more about exercising power *with* rather than *over* others, holding out the prospect of positive gains that are too great to pass up. Positive-sum gains, however, do not equate to equal-sum gains and some gains turn out to be illusory, as noted, and in this way skewed reward structures may be 'masked' by the exercise of quieter registers of power: through persuasion, inducement and manipulation. On this view, the 'power to' engineer financial solutions to their advantage is an apt description of what financial intermediaries attempt to do. And, in the last resort, it is what they do that is said to matter.

Conclusion

In contrasting a broadly structural account of power, where the City reproduces itself as a dominant force over time, to a more provisional, expedient landscape of the City's achievements, I inevitably run the risk of overstating the difference. But the difference, even if drawn a little too starkly, rests upon divergent assumptions about how finance and the UK economy is organised and how open it is to change. How we 'read' the basis of the City's power matters because it spells out the room for manoeuvre that there is to see things differently.

If the City's dominance is more or less systemic, then the argument for a more regulated, less lucrative, financial sector is subject to structural constraints that may well reassert themselves after several lean years. Derivatives and securitisation are, after all, not processes that can simply be wished away and nor necessarily should they be. A more provisional account of the City's powers, however, one which emphasises the 'produced' nature of those powers under contingent circumstances, offers greater room for manoeuvre about what state regulation can achieve and what role the City and finance should assume in the UK economy going forward. This is not to imply some kind of voluntaristic action directed at the worst excesses of financialisation, but rather that such a reading opens up more of a space for political engagement. An engagement that is about what kind of economy the UK should be if, as is now evident, the notion that the country could survive on finance alone has gone the way of much of the City's speculative gains pocketed in recent years.

Further reading

- For an 'insiders' view of the City's growth and future (just prior to the global financial crisis), see Gieve (2007).
- Massey (2007) offers a more critical account of the rise of the City and its distortion of the UK economy.
- Michie (1992) provides an historical account of the City's dominance.

References

Allen, J. (2008) Pragmatism and power, or the power to make a difference in a radically contingent world, *Geoforum*, 39: 1613–24.

Allen, J. (forthcoming.) Powerful city networks: more than connections, less than domination and control, *Urban Studies*.

Blackburn, R. (2006) Finance and the fourth dimension, *New Left Review*, 39: 39–70.

Blackburn, R. (2008) The subprime crisis, *New Left Review*, 50: 63–106.

Bourdieu, P. (1989) Social space and symbolic power, *Sociological Theory*, 7: 14–25.

Cain, P.J. and Hopkins, A.G. (1987) Gentlemanly capitalism and British expansion overseas II, *Economic History Review*, XL, 1: 1–26.

Gieve, J. (2007) The City's growth: the crest of a wave or swimming with the stream, *Bank of England Quarterly Bulletin*, Q2: 286–90.

Harris, L. (1988) Alternative perspective on the financial system, in L. Harris, J. Coakley, M. Croasdale and T. Evans, (eds), *New Perspectives on the Financial System*. Beckenham: Croom Helm.

Ingham, G. (1984) *Capitalism Divided? The City and Industry in British Social Development*. Basingstoke: MacMillan.

Ingham, G. (1989) *Commercial Capital and British Development*: a reply to Michael Barratt Brown, *New Left Review*, 172: 45–65.

Lee, C.H. (1986) *The British Economy Since 1700: A Macroeconomic Perspective*. Cambridge: Cambridge University Press.

Leyshon, A. and Thrift, N. (2007) The capitalisation of almost everything: the future of finance and capitalism, *Theory, Culture and Society*, 24: 97–115.

Massey, D. (2007) *World City*. Cambridge: Polity.

Michie, R.C. (1992) *The City of London: Continuity and Change 1850–1990*. Basingstoke: Macmillan.

Pryke, M. and Allen, J. (2000) Monetized time-space: derivatives – money's 'new imaginary', *Economy and Society*, 29: 264–84.

Sassen, S. (2009) Cities in today's global age, *SAIS Review*, XXIX: 3–34.

Savage, M. and Williams, K. (eds) (2008) *Remembering Elites*. Oxford: Sociological Review Monographs.

Thrift, N. (1994) On the social and cultural determinants of international financial centres: the case of the City of London, in S. Corbridge, R. Martin and N. Thrift (eds), *Money, Power and Space*. Oxford: Blackwell.

5

BANKING ON FINANCIAL SERVICES

Shaun French, Karen Lai and Andrew Leyshon

AIMS

- To account for the growing importance of the financial services sector to the UK economy

- To explain the uneven geography of financial services employment and the concentration of financial services in particular cities

- To map the changing geography of UK financial centres and identify 'winners' and 'losers' in the spatial competition to attract financial jobs

Since at least the 1980s both London's financial district and the financial services sector more generally have become increasingly important to the UK economy (and see Chapter 4). The role of financial services as a key engine of economic growth has led recent writers to argue that the UK should now be considered a financialised economy (Erturk et al., 2008); that is, one in which the corporate sphere is increasingly dominated by the demands of financial markets, and the economy is increasingly dependent upon the financial services industry to sustain growth.

The most visible spatial manifestation of the financialisation of the UK economy has been the growing significance and prestige of financial centres. Following a long period of relative decline, the late 1970s and 1980s not only witnessed a renaissance in the City of London's role as a premier international financial centre, but also the resurgence of the fortunes of many provincial centres (Leyshon et al., 1989). Large provincial cities, such as Edinburgh, Manchester, Bristol and Leeds, have competed with one another to attract financial firms, and have sought to position themselves as leading financial centres on the regional, national and international stage (see Figures 5.1 and 5.2). For these cities, their reinvention as financial centres has been central to job creation efforts and city centre regeneration strategies in the wake of earlier episodes

Figure 5.1 Marketing material from the mid-1990s promoting Bristol as the second largest financial centre outside of London
Source: Unknown

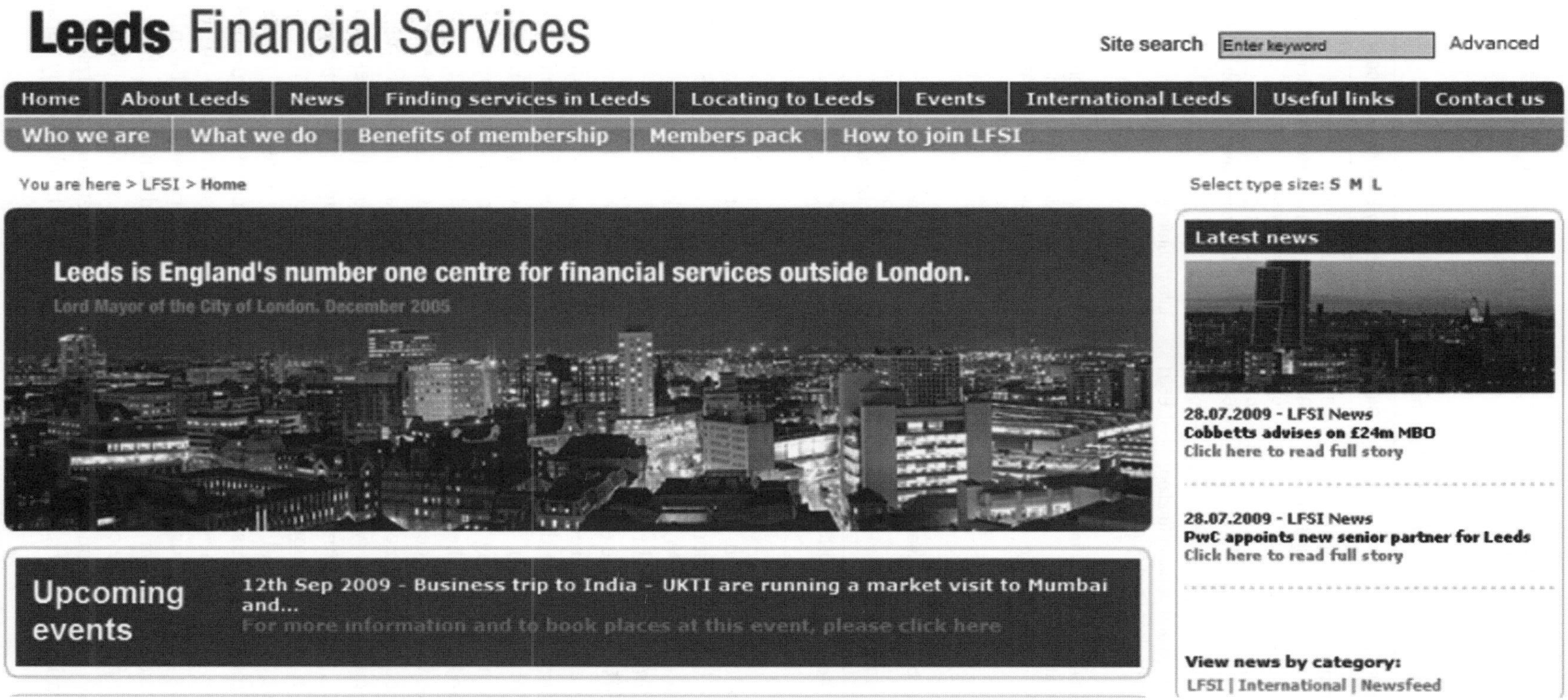

Leeds Financial Services

Site search [Enter keyword]   Advanced

| Home | About Leeds | News | Finding services in Leeds | Locating to Leeds | Events | International Leeds | Useful links | Contact us |
| Who we are | What we do | Benefits of membership | Members pack | How to join LFSI |

You are here > LFSI > Home

Leeds is England's number one centre for financial services outside London.

Lord Mayor of the City of London, December 2005

Select type size: S M L

Latest news

28.07.2009 - LFSI News
Cobbetts advises on £24m MBO
Click here to read full story

28.07.2009 - LFSI News
PwC appoints new senior partner for Leeds
Click here to read full story

View news by category:
LFSI | International | Newsfeed

Upcoming events

12th Sep 2009 – Business trip to India – UKTI are running a market visit to Mumbai and...

For more information and to book places at this event, please click here

Figure 5.2 Leeds Financial Services Initiative website, summer 2009 (www.leedsfinancialservices.org.uk, reproduced with kind permission of LSFI)

Table 5.1   Definition of financial services

SIC category	SIC no.	SIC activity description (2003)
65. Financial intermediation, except insurance and pension funding	6511	Central banking
	6512	Other monetary intermediation
	6521	Financial leasing
	6522	Other credit granting
	6523	Other financial intermediation not elsewhere classified
66. Insurance and pension funding, except compulsory social insurance	6601	Life insurance
	6602	Pension funding
	6603	Non-life insurance
67. Activities auxiliary to financial intermediation	6711	Administration of financial markets
	6712	Security broking and fund management
	6713	Activities auxiliary to financial intermediation not elsewhere classified
	6720	Activities auxiliary to insurance and pension funding
74. Other business activities	7411	Legal activities
	7412	Accounting, book-keeping and auditing activities; tax consultancy

Source: Adapted from ONS (http://www.statistics.gov.uk)

of deindustrialisation. However, the severity and extent of the financial crisis of 2007–8 has raised serious questions over the future of a finance-led model of economic growth (French et al., 2009), not only for the UK as a whole, but for individual cities and regions.

In this chapter we examine the changing geography of UK financial centres. Our analysis is based on an examination of Office for National Statistics' (ONS) employment data, as well as previous studies of financial centres. In terms of the former, financial services are broadly defined to include activities such as banking and insurance, grouped under the heading of financial intermediation in the ONS Standard Industry Classification (SIC), as well as legal services and accountancy (Table 5.1). This is for two reasons: first, much of the commercial activity conducted by legal and accounting firms complements and overlaps with the services offered by financial services companies; second, previous analyses have made clear the significant role played by these in the development of provincial financial centres over the past two decades (French and Leyshon, 2003; Leyshon et al., 1989). The remainder of the chapter is organised as follows. In the second section we begin by looking in more detail at the factors behind the growing significance of the financial services industry during the last quarter of a century or so and explanations for the concentration of financial services in financial centres. In the third section we map out the changing geography of financial services employment and identify 'winners' and 'losers' in the spatial competition to attract financial corporations and jobs. The fourth section concludes the chapter by considering the likely impact of the 2007–8 financial crisis on the future geography of UK financial centres.

Financial centre dynamics

Before looking in detail at the contemporary geography of financial centres, it is important to provide a brief overview of the history of financial services in the UK and of the factors that influence the growth and dynamics of financial centres. Financial services employment has grown markedly over the past two decades. In 1984 the financial services sector accounted for just over 1 million full-time equivalent (FTE) jobs (FTE employment is calculated by adding half the number of part-time to the total number of full-time jobs); by 2006 employment had grown to approximately 1.4 million FTE jobs, an increase of 36 per cent. During this period not only was there an absolute increase in employment in the sector, but the contribution of financial services to national employment grew from 5.55 per cent to 6.30 per cent. As well as being an increasingly important source of employment, the financial services industry also contributes significantly to UK gross domestic product (GDP) and tax receipts. Financial services employees accounted for some 13.0 per cent of total UK income tax revenue in 2006, and financial services firms contributed more than a quarter (27.0 per cent) of all UK corporation tax (IFSL, 2008). Furthermore, in 2006 the financial services industry's share of UK GDP amounted to 7.7 per cent (IFSL, 2008). Despite the ravages of the financial crisis, bank assets were still equivalent to over 400 per cent of GDP by the end of 2008 (*Economist*, 2009).

The growth in importance of financial services can in part be understood as an outcome of a policy of competitive re-regulation of both wholesale markets (international financial markets and commercial finance activities that are chiefly located in London) and retail markets (personal and high-street financial services) by the Conservative government during the 1980s. The system of structural regulation that had prevailed for much of the post-war period was replaced by a new regime of prudential regulation. For retail financial services markets, the Building Societies Act and the Financial Services Act (both passed in 1986) made it possible for financial services firms to compete much more freely with one another in what were previously strictly circumscribed product markets. It has now become commonplace, for example, for banks to offer mortgages, the traditional raison d'être of building societies, and for building societies, conversely, to offer current accounts. Rather than guard against financial instability through the application of strict rules on what products different types of financial institution were allowed to offer, and in what markets they are allowed to do business, henceforth the stability of the financial system was to be ensured by firms demonstrating compliance with regulatory requirements for prudential financial management. These requirements largely took the form of rules governing the size of financial reserves banks, building societies and insurance companies must hold. While the system of prudential regulation has recently come under serious scrutiny following the government rescue of Northern Rock, Royal Bank of Scotland and Lloyds-HBOS in 2007–8, the increase in competition that occurred following its introduction spurred financial innovation. Notable innovations included the development of new retail financial services markets for sub-prime and buy-to-let mortgages, the repackaging or securitisation of consumer debt to be traded in the secondary banking markets, as well as the development of new delivery channels such as telephone and Internet banking.

Processes of financial innovation, securitisation and re-regulation have, in turn, driven the increasing financialisation of corporate and everyday life. In the case of the former it has been argued that there has been a significant shift in the strategic orientation of corporations, whereby competition in product markets has become subsidiary to the principal aim of securing 'shareholder value' to satisfy the demands of financial markets (Erturk et al., 2008). Everyday life is also argued to have become financialised as households and individuals are increasingly compelled to make provision for their own long-term financial futures in the face of the retreat of the welfare state and the demise of final salary occupational pension schemes (Langley, 2008). In turn, the financialisation of everyday life, underwritten by the growing importance governments have come to place on market solutions and the need for individuals to exercise greater personal responsibility, has significantly increased the demand for financial services.

However, the economic benefits accrued from the rapid expansion of the financial services industry have been highly geographically uneven (see also Chapter 3). Firms have exhibited a strong tendency to cluster in select locations and, as such, a large proportion of financial activity is concentrated within a relatively small group of urban financial centres, chief amongst which is the London financial district. In 2006, London accounted for 44 per cent of the value added to UK GDP by the financial services sector (IFSL, 2008), and over one-quarter (26.5 per cent) of total financial services employment. Moreover, the top 10 financial centres measured together accounted for some 46 per cent of all financial services employment in the UK in the same year (Table 5.2). While the dominance of the City of London – which is one of the largest and most important financial centres in the world – is unlikely to be seriously threatened in the foreseeable future, its very success has brought its own set of problems such as congestion and high labour and property costs, which have in turn provided the stimulus for the decentralisation of financial activity from the capital. During the 1970s and 1980s diseconomies of agglomeration, coupled with the City of London's growing focus on the international market in the wake of the 'Big Bang' reforms of the stock market in 1986, led to the relocation of much back office and retail financial services activity away from London and into the wider South-east region, as well as along the M4 corridor into the South-west and further afield (Leyshon et al., 1989).

Table 5.2 Top 20 financial centres ranked by employment (LAD).

Ranking	1984[a]		1991[a]		1995[b]		2000[b]		2006[c]	
	LAD	FTE	LAD	FTE	LAD	FTE	LAD	FTE	LAD	FTE
1	London	263754	London	301634	London	315856	London	381366	London	370329
2	Birmingham	30917	Birmingham	35654	Birmingham	39129	Birmingham	42288	Manchester	47171
3	Manchester	28776	Manchester	35213	Edinburgh	34607	Manchester	40414	Leeds	38688
4	Glasgow	23290	Edinburgh	31879	Manchester	34419	Edinburgh	39519	Edinburgh	37771
5	Edinburgh	20131	Bristol	30567	Bristol	32025	Glasgow	32125	Birmingham	35993
6	Bristol	19714	Glasgow	24684	Leeds	27680	Leeds	30213	Bristol	30707
7	Leeds	15968	Leeds	24525	Glasgow	25611	Bristol	29535	Glasgow	29831
8	Liverpool	13840	Liverpool	17021	Liverpool	15804	Liverpool	14951	Liverpool	18710
9	Croydon	11689	Norwich	14604	Croydon	12439	Cardiff	12904	Cardiff	16838
10	Brighton and Hove	9758	Croydon	13797	Cardiff	11867	Sheffield	12726	Sheffield	15052
11	Norwich	9179	Brighton and Hove	12617	Sheffield	11200	Brighton and Hove	12153	Newcastle-upon-Tyne	14674
12	Sheffield	9166	Newcastle-upon-Tyne	11423	Norwich	11056	Croydon	11280	Norwich	14598
13	Newcastle-upon-Tyne	9010	Cardiff	11170	Bradford	10934	Calderdale	11108	Nottingham	11077
14	Cardiff	8774	Reading	11101	Calderdale	10435	Norwich	11097	Bradford	10399
15	Sefton	8686	Sheffield	10710	Newcastle-upon-Tyne	10036	Reading	10676	Reigate and Banstead	9946
16	Bradford	7760	Bradford	9555	Bournemouth	9437	Newcastle-upon-Tyne	10463	Bournemouth	9887
17	Southend-on-Sea	7159	Northampton	9509	Nottingham	9283	Southampton	10224	Brighton and Hove	9479
18	Leicester	7131	Nottingham	9198	Northampton	9016	Bradford	9449	Chester	9472
19	Reading	7081	Southampton	9075	Reading	8933	Leicester	8808	Calderdale	9202
20	Nottingham	6937	Bournemouth	8690	Southampton	8501	Bournemouth	8773	Bromley	9019

Notes: Annual Business Inquiry data from NOMIS. Ranking based on full-time equivalent (FTE) employment (total number of full-time workers and half of part-time workers). Local authority district (LAD) boundaries used except for London which is Greater London, Birmingham includes Solihull, Bristol includes South Gloucestershire, and Manchester includes Salford and Trafford.

a Based on SIC 8140, 8150, 8200, 8310, 8320, 8350, 8360 (1980 definition)

b Based on SIC 65, 66, 67, 7411, 7412 (1992 definition)

c Based on SIC 65, 66, 67, 7411, 7412 (2003 definition)

Source: Author's own analysis of Annual Business Inquiry employment data.

Explanations of why financial services firms cluster in particular cities and understandings of the drivers that generate change in the hierarchy of financial centres focus on four factors (Bailey and French, 2005; French and Leyshon, 2003; Leyshon et al., 1989).

- *Access or proximity to market:* Demand for many types of financial service is closely correlated to concentrations of population and businesses. One important reason why firms locate in large urban centres – and which also explains why financial centre hierarchies often closely mirror that of the existing urban hierarchy – is the desire to access local and regional markets. Changes in local market conditions can act as powerful push or pull factors, attracting firms to centres in which demand is growing or, conversely, driving firms away from centres in which demand is falling or has become saturated. Leyshon et al. (1989) argued that an increase in demand for financial services in many regional cities during the 1980s combined with processes of decentralisation to bring about the resurgence of provincial financial centres. In the case of the growth of Bristol and Leeds, research has emphasised the role of local demand and the ability of local firms to compete with financial services providers in the City of London, particularly in the provision of corporate finance services which enable local small- and medium-sized companies to access capital through stock markets (Bailey and French, 2005; DTI, 2001: 45; French and Leyshon, 2003). Conversely, smaller subregional centres, such as Nottingham for example, have struggled to expand and develop as a result of the limited extent of local markets and the danger that firms will choose to relocate to larger and more successful regional centres – such as Birmingham in the case of Nottingham – in order to access larger markets (Mincher and Nettleship, 2007).

- *Urban infrastructure:* Firms are attracted to urban centres not only because of local demand but also because cities have the infrastructure to supply firms with the assets they need to produce financial services, including skilled labour, office space and information communications networks. While larger cities are generally better placed to supply such assets than centres further down the urban hierarchy, significant variance exists between the types of assets that cities are able to supply. One of the great attractions of the City of London for financial services firms is precisely the access it provides to a highly skilled and specialised pool of financial labour, to a supply of commercial office space tailored to the needs of financial firms competing in international markets, as well as offering the prestige of a City of London address (Corporation of London, 2003). Nevertheless, while London's status as a global city means that it enjoys a significant advantage over its competitor centres in the UK, the success of London's financial district has, in turn, placed new strains on its existing urban infrastructure, increasing the cost of property and labour and enabling regional centres to position themselves as attractive alternative locations, especially for back-office functions. More generally, factors such as the nature of the local commercial property and labour markets and the creation of a credible financial centre identity are key means by which cities such as Edinburgh, Bristol and Leeds seek to compete with one another to attract financial firms.

- *Internal economies of scale and scope:* These are the benefits that can accrue from the concentration of activities within a single firm or site. While internal economies do not on their own explain the tendency for financial services firms to cluster in specific centres, they can operate as a powerful force for reinforcing the concentration of financial services activity, especially when considered in conjunction with the key role of cities in supplying the infrastructure (principally property and labour) that enable firms to take advantage of potential economies of scale and scope. Changes in market conditions, the application of new technology – such as the introduction of telephone distribution channels, the emergence of call centres, or pressures to increase the efficiency of internal economies through mergers and acquisitions – can all threaten the existing spatial hierarchy, presenting opportunities and threats for individual financial centres (Bailey and French, 2005). On the one hand, centres such as Newcastle have benefited considerably from the new opportunities to reap internal economies afforded by call centres (Richardson et al., 2000). The centralisation of customer services in call centres enables cost savings and the opportunity to significantly shrink high-street branch networks (Leyshon et al., 2008). On the other hand, the dependence of many smaller, sub-regional centres on a small number of large financial services employers renders some cities vulnerable to the future pursuit of new internal economies (Mincher and Nettleship, 2007). Such cost and efficiency savings could be sought through concentrating functions located at multiple sites onto a single site, or as a result of the merger/acquisition of one firm by another, such as the recent takeover of HBOS by Lloyds.

- *Agglomeration economies*: It is important to distinguish between processes of agglomeration on the one hand and the co-location of financial services on the other. While forces of demand, urban infrastructure and internal economies of scale and scope frequently lead firms to co-locate, it is often the case that there exists very little interaction between firms that are located in the same city-region. In contrast, financial services agglomerations or clusters, as epitomised by the likes of the City of London and Wall Street in New York, are characterised by dense networks of inter-firm trade, professional contacts and knowledge interdependencies (Faulconbridge et al., 2007). Not only do such interdependencies frequently lead firms to cluster within a relatively small geographical area, but the significant advantages that can be accrued from the external economies of complexity, scale and scope, as well as the ability to tap into vital knowledge networks and epistemic communities represents a powerful pull factor for many financial services firms (Faulconbridge, 2007;

French, 2002). In the case of the City of London, for example, firms have traditionally clustered in the square mile area around the Bank of England in order to maximise the advantages of agglomeration and what Storper (1997) has termed 'traded and untraded interdependencies'. While it is possible to draw a clear conceptual distinction between agglomeration and co-location, in practice a financial centre can often exhibit characteristics of both. As in the case of the three other factors, agglomeration economies can act as both a pull factor in attracting firms to a financial centre, as well as acting, in certain instances, to push firms away from a specific centre. The high concentration of activity in the City of London has, as discussed above, significantly driven up the costs of doing business and has led many retail financial firms to relocate out of central London. Such periodic waves of decentralisation clearly offer opportunities for other UK cities to attract firms looking for cheaper locations.

Financial centre geographies

Having identified some of the key reasons behind the growing contribution of the financial services sector to the UK economy, as well as explanations for the geographical concentration and clustering of financial services, we will now examine the changing geography of UK financial centres. There are two major aspects to this.

The first is trends in employment and rankings. Table 5.2 ranks the top 20 Local Authority Districts (LAD) in Britain with the highest number of financial services employees from 1984 through to 2006. The table highlights the continued pre-eminence of London and the relative stability of financial centres in the top seven positions (London, Manchester, Leeds, Edinburgh, Birmingham, Bristol and Glasgow). The geography of financial services employment is therefore made up of a 'core' of financial centres. However, the success of each individual city in attracting and retaining financial services is likely to be the consequence of different combinations of the four factors discussed above.

While there is a well-established framework of financial centres at the top of the hierarchy, it is also possible to identify winners and losers. For example, despite retaining its status as eighth largest centre, Liverpool's growth has trailed noticeably behind that of the larger centres. In contrast, Leeds has benefited significantly from financialisation, successfully transforming itself into one of the largest regional financial centres. Cardiff is another city that has moved up the rankings and more than doubled its FTE employment between 1984 and 2006. In addition (and despite being absent from Table 5.2 because crown dependencies produce their own independent employment statistics), Jersey, Guernsey and the Isle of Man have all also managed to successfully reinvent themselves as 'off-shore' financial centres. The most recent global financial centre index compiled by the City of London corporation goes as far as to rank Guernsey (12th), Jersey (13th) and the Isle of Man (18th) as the most important UK financial centres after London (1st) (Yeandle et al., 2009). The next most important international centre based in the UK is Edinburgh (20th). However, solely in terms of numbers employed, Guernsey (7,164 financial services employees in 2006 (States of Guernsey, 2008)), Jersey (11,420 employees in 2006 (States of Jersey, 2008)), and the Isle of Man (9,395 employees in 2006 (Isle of Man Treasury, 2008)) are comparable to sub-regional centres like Nottingham.

In the case of the smaller centres more generally, Table 5.2 suggests that the gap between, on the one hand, London and the regional centres and, on the other hand, the sub-regional centres (the largest of which is Liverpool) has progressively widened since the 1980s. Nevertheless a number of smaller cities and towns have clearly benefited, albeit in more modest terms, from the growth of financial services. The likes of Nottingham, Chester and Bournemouth have all moved up the rankings since the mid-1980s. These smaller centres have grown as the result of either the relocation of back-office functions or the creation of new financial products and services with increased demand for office space, such as call centres and credit card centres. More specifically, Nottingham has benefited from the establishment of significant operations in the city by Capital One and Experían, while Bournemouth and Chester have benefited, respectively, from the decision of large US credit card firms American Express and MBNA to establish significant European operations in each location.

The second aspect to financial centre geographies concerns trends in relation to co-location and agglomeration. In addition to the number of financial services employees, the relative significance of financial centres can also be assessed in terms of the existence or otherwise of agglomerative economies. For many commentators, agglomerative economies are an essential characteristic of a true financial centre, not least because they are seen to confer critical competitive advantages on firms, particularly with regard to innovation. Nevertheless, it is notoriously difficult to identify and measure clustering or agglomerative economies (French, 2002). By analysing and comparing financial employment data at both the LAD scale (which provides a measure of employment in the city centre) and the larger scale of the travel-to-work area or TTWA (which provides a measure of employment in the wider city-region), an indication of the relative geographical concentration and thus the relative agglomeration of financial activity in a particular centre can be gauged. Figure 5.3 plots a best-fit line of correlation for employment at the TTWA and LAD scales for leading UK financial centres (excluding London). Centres that appear above the line have proportionately more employment within the LAD than the TTWA, illustrating a greater spatial concentration of financial activity. Conversely, centres that fall below the line have proportionately more employment within the TTWA than in the LAD, suggesting that their financial services employment is less spatially concentrated and more likely to be co-located rather than agglomerated. Centres that emerge as having particularly agglomerated financial services employment include

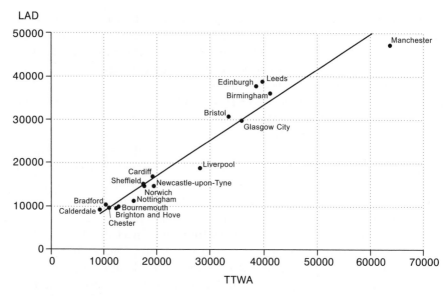

Figure 5.3 Correlation of financial services employment for leading UK financial centres (excluding London)
Source: Authors' own analysis of Annual Business Inquiry employment data.

Leeds, Edinburgh, Birmingham and Bristol. In contrast, other urban centres such as Manchester, Liverpool and Nottingham fall below the best-fit line, suggesting that a significant proportion of financial services employment in these centres is likely to be co-located rather than agglomerated.

While the comparison of TTWA and LAD data provides a starting point for distinguishing between concentrations of financial services activity and true clusters, previous studies have sought to develop more sophisticated quantitative measures. One such study, carried out in 2001 by the Department of Trade and Industry (DTI; the forerunner of the current Department of Business, Innovation and Skills), sought to distinguish 'true' clusters through the application of a statistical model involving the analysis of financial services employment data relative to resident population, to total employment, and to the relative concentration of financial services business units and large employers. Another report by the Financial Services Skills Council (FSSC) in 2008 builds on the insights of the DTI research. Both studies underlined the status of London as the UK's premier financial services cluster, with the DTI concluding that 'at best some of the regional clusters have only weak cluster characteristics compared to the City' (DTI, 2001: 45). More specifically, the DTI report differentiated between strong regional financial services clusters in Edinburgh (fund management and banking), Leeds and Bradford (building societies, non-life insurance and consumer credit) and Bristol (life insurance and building societies), and less substantial clusters with more limited evidence of agglomerative economies in cities such as Liverpool, Manchester and Norwich. The more recent study by the FSSC (2008) identified seven

Table 5.3 Financial services clusters

Location	Stage	Depth	Employment	Significance
City of London*	Mature	Deep	Growing	International
Canary Wharf	Emerging	Deep	Growing	International
Bristol	Established	Shallow	Stable	National
Norwich	Established	Deep	Growing	National
Manchester**	Established	Shallow	Growing	National/International
Leeds	Established	Shallow	Growing	National/International
Edinburgh	Established	Deep	Growing	International
Glasgow	Established	Shallow	Growing	National/International

Notes: *Includes Westminster and City.
**Includes Macclesfield and Stockport.
Source: Adopted from FSSC, 2008

financial services clusters (Table 5.3), assuming that the City of London and Canary Wharf are counted as one cluster, and, in contrast to the DTI, the FSSC report identifies Glasgow, Manchester and Norwich as true clusters. In addition, smaller-scale, qualitative studies of particular centres have also been undertaken, with research on Edinburgh and Bristol revealing evidence of agglomerative processes (Bailey and French, 2005; French, 2002).

Place marketing, and financial centre strengths and weaknesses

The location quotient (LQ) is a measure of the ratio of an area's share of employment benchmarked against national figures. LQ analysis has been applied here to reveal how different financial centres have fared over the years in terms of their share of employment in particular financial service sectors compared to national level data, and acts as a useful barometer of financial strengths and weaknesses. An LQ of higher than 1.0 means it has a higher share of employment in particular industries or sub-sectors compared to other financial centres. Table 5.4 shows the LQs for the financial services sector for the largest seven centres for 1991 and 2006. It reveals that the leading financial centres identified earlier all have LQs over 1.0 and that, surprisingly, Edinburgh emerges above London as the 'most financialised' with the highest share of financial services employment as a proportion of total employment. Detailed LQs for constituent financial services sub-sectors for London and Edinburgh are shown in Figures 5.4 and 5.5. Figure 5.4 illustrates the depth and strength of London as a financial centre, with high LQ indices for most of the sub-sectors of the industry. However, it also illustrates that by 2006 London had lost its pre-eminence in *all* financial services sectors, as the LQ for the other credit granting and the life insurance industry fell below 1.0, indicating that the capital's share of employment in these sectors was less than expected. The declining importance of these sectors within the

Table 5.4 LQ of the financial services sector for the largest financial centres

City	1991	2006
Edinburgh	1.64	2.03
London	1.96	1.86
Leeds	1.34	1.58
Bristol	1.61	1.42
Manchester	1.03	1.30
Glasgow	0.91	1.16
Birmingham	0.94	1.04

Source: Authors' own analysis of Annual Business Inquiry employment data.

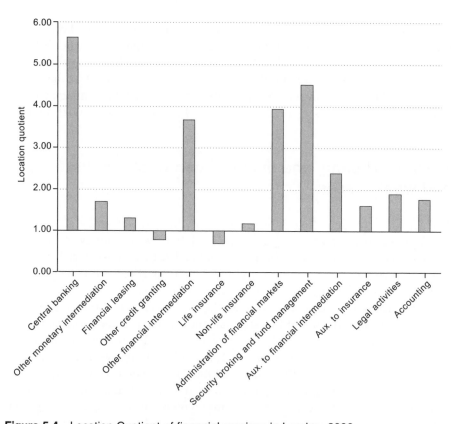

Figure 5.4 Location Quotient of financial services in London, 2006
Source: Authors' own analysis of Annual Business Inquiry employment data.

London financial district is no doubt a result of the diseconomies of agglomeration referred to earlier.

While no other financial centre in the UK demonstrates London's all-round strength, many of the regional centres display particular strengths within

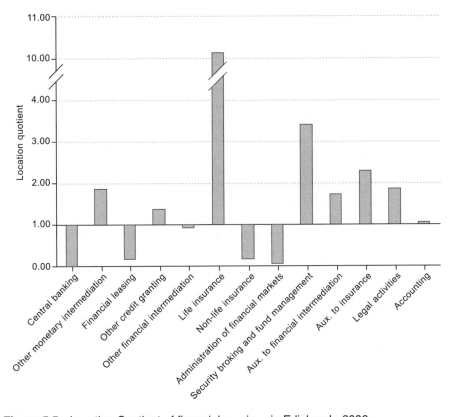

Figure 5.5 Location Quotient of financial services in Edinburgh, 2006.
Source: Authors' own analysis of Annual Business Inquiry employment data.

parts of the overall industry. Figure 5.5 illustrates, for example, that whereas London is characterised by a relative weakness in terms of life insurance, Edinburgh is by contrast particularly strong in this area. As well as being home to large life insurance companies like Scottish Widows and Standard Life, Edinburgh is also second only to London in the UK in terms of the brokering and fund management industry (notable Edinburgh fund managers include firms such as Martin Currie and Baillie Gifford). Edinburgh has traditionally also been strong in the area of banking, as exemplified by the headquarters in the city of Royal Bank of Scotland and HBOS, prior to its takeover by Lloyds. Similarly, the presence of firms such as AXA Life and Clerical & Medical attest to the strength of the life assurance sector in Bristol, and Direct Line and Leeds Building Society to Leeds' particular strengths in the non-life insurance and building society sectors (Bailey and French, 2005; French, 2002; FSSC, 2008). Other areas of specialisation include private banking in Jersey, captive insurance in Guernsey, and life insurance in the Isle of Man (Cobb, 1998).

As the contribution of financial services to urban and regional economies has grown, so have parallel efforts at marketing. To this end many local authorities have established partnerships with the local financial services industry in order to promote and market themselves as financial centres, as well as encourage the sort of networking and sharing of expertise considered to be the hallmark of a successful cluster. While London, through the high-profile work of bodies such as TheCityUK and International Financial Services London (IFSL), is at the forefront of such efforts, other cities have become increasingly active. Leeds Financial Services Initiative (LFSI), established in 1993, represents a particularly good example (Figure 5.2), and similar bodies exist to promote Manchester (Pro-manchester), Liverpool (Professionaliverpool) and Jersey (Jersey Finance), while Glasgow has established a purpose-built International Financial Services District (www.ifsdglasgow.co.uk) in order to attract financial firms and raise its profile on the international stage. As with agglomerative economies, it is difficult to measure the impact of such initiatives. However, in the case of Jersey, Guernsey and the Isle of Man there is evidence to suggest that these off-shore centres have been particularly successful at generating a 'reputational advantage', whereby perceptions about these centres are higher than indicated by purely quantitative measures such as the cost and availability of skilled labour, and tax policies (Yeandle et al., 2009: 37).

Conclusions: after the crisis

In this chapter we have sought to account for the growing importance of the financial services sector to the UK economy, and have shown how much of this activity is concentrated in a relatively small number of financial centres, at the apex of which sits the City of London. Cities such as Edinburgh, Leeds, Manchester and Bristol have benefitted considerably from the growth of the financial services industry, repositioning themselves as leading financial centres. However, as suggested at the beginning of the chapter, the unprecedented scale and scope of the 2007–8 financial crisis represented a significant challenge to many of these cities. The crisis dealt a serious blow to the credibility and stability of the UK financial system, not only calling into question prevailing business models, but also leading to the nationalisation of some of the UK's most prominent banks.

The immediate aftermath of the crisis witnessed a significant downturn in the sector and a swathe of financial job losses. While initially centred on London and on the investment banking sector in particular, exemplified by high-profile redundancies at the City offices of firms such as Lehman Brothers, Merrill Lynch and UBS, the downturn rapidly engulfed other UK financial centres. There were far-reaching job cuts in all three banks rescued by the Government, with the Lloyds banking group (including HBOS) announcing more than 8,000 redundancies in the first half of 2009 alone at sites in Bristol, Scotland, West Yorkshire and Kent. While by no means restricted to head office functions, such cuts had a detrimental impact on the status and reputation of Edinburgh, with the forced rescue of Royal Bank of

Scotland and HBOS effectively signalling the end of Edinburgh's role as an independent banking centre, while Newcastle's status as a sub-regional financial centre suffered following the collapse of Northern Rock. Meanwhile, the thinning out of capacity and institutions in the City of London as a result of the crisis not only reduced competition in many markets and facilitated the rapid recovery of profits in surviving firms, but also reduced some of the diseconomies of agglomeration that helped fuel the spatial dispersion of many financial activities to regional financial centres. Indeed, the unprecedented amounts of government money spent on shoring up the balance sheets of UK banks – estimated to be in the order of £1.3 trillion in the form of special loans, investment, guarantees and the creation of new money – can be seen as yet another example of 'accidental regional development policy' that has favoured more affluent regions of the UK (in this case London and the Southeast) over more disadvantaged regions. As such, there are echoes of the ways in which defence spending favoured already affluent regions in the 1970s and 1980s (Lovering, 1985). While most commentators agree that government intervention was necessary to support the balance sheets of banks in order to protect the viability of the wider financial system and the wider national and global economy, the stimulus package disproportionately benefitted the London economy. Therefore, while London might have appeared as the initial loser in the crisis, it's dominance and importance ensured that it is likely to emerge as the major beneficiary among UK financial centres.

Further reading

- Erturk et al. (2008) provides an excellent overview of work on financialisation.
- Tickell (2000) explains the uneven geographies of finance and their impact on regional development.
- The edited volume by Martin (1999) provides further insights into the geography of financial services in the UK.

References

Bailey, N. and French, S. (2005) The locational dynamics of financial services in regional cities, in N. Buck, I. Gordon, A. Harding and I. Turok (eds.), *Changing Cities: Rethinking Urban Competitiveness. Cohesion and Governance.* Basingstoke: Palgrave Macmillan. pp. 112–31.

Cobb, S.C. (1998) Global finance and the growth of offshore financial centers: The Manx experience, *Geoforum,* 29: 7–21.

Corporation of London (2003) *Financial Services Clustering and its Significance for London.* London: The Corporation of London.

DTI (2001) *Business Clusters in the UK: A First Assessment.* London: Department of Trade and Industry.

Economist, The (2009) Regulating banks: The devil's punchbowl. 9 July.

Erturk, I., Froud, J., Johal, S., Leaver, A. and Williams, K. (eds) (2008) *Financialisation at Work: Key Texts and Commentary.* London: Routledge.

Faulconbridge, J.R. (2007) Exploring the role of professional associations in collective learning in London and New York's advertising and law professional-service-firm clusters, *Environment and Planning A,* 39: 965–84.

Faulconbridge, J., Engelen, E., Hoyler, M. and Beaverstock, J. (2007) Analysing the changing landscape of European financial centres: The role of financial products and the case of Amsterdam, *Growth and Change,* 38: 279–303.

French, S. (2002) Gamekeepers and gamekeeping: assuring Bristol's place within life underwriting, *Environment and Planning A,* 34: 513–41.

French, S. and Leyshon, A. (2003) City of Money? in M. Boddy (ed.), *Urban Transformation and Urban Governance.* Bristol: Policy Press. pp. 32–51.

French, S., Leyshon, A. and Thrift, N. (2009) A very geographical crisis: the making and breaking of the 2007–2008 financial crisis, *Cambridge Journal of Regions, Economy & Society,* 2: 287–302.

FSSC (2008) *Financial Services Clusters.* London: Financial Services Skills Council.

IFSL (2008) *Economic Contribution of UK Financial Services 2008.* London: International Financial Services London.

Isle of Man Treasury (2008) *2008 Digest of Economic & Social Statistics.* Douglas: Isle of Man Government.

Langley, P. (2008) *The Everyday Life of Global Finance.* Oxford: Oxford University Press.

Leyshon, A., French, S. and Signoretta, P. (2008) Financial exclusion and the geography of bank and building society branch closure in Britain, *Transactions of the Institute of British Geographers,* 33: 447–65.

Leyshon, A., Thrift, N. and Tommey, C. (1989) The rise of the British provincial financial centre, *Progress in Planning,* 31: 312–41.

Lovering, J. (1985) Regional intervention, defense industries, and the structuring of space in Britain: the case of Bristol and South-Wales, *Environment and Planning D: Society and Space,* 3: 85–107.

Martin, R. (1999) *Money and the Space Economy.* Chichester: Wiley.

Mincher, H. and Nettleship, K. (2007) *Nottingham: International Financial Services Gateway to the Future.* Nottingham: Nottingham City Council.

Richardson, R., Belt, V. and Marshall, N. (2000) Taking calls to Newcastle: the regional implications of the growth in call centres, *Regional Studies,* 34: 357–69.

States of Guernsey (2008) *Guernsey Facts and Figures 2008.* St Peter Port: States of Guernsey.

States of Jersey (2008) *Jersey Economic Digest 2008.* St Helier: States of Jersey.

Storper, M. (1997) *The Regional World.* New York: Guilford Press.

Tickell, A. (2000) Finance and localities, in G. Clark, M. Feldman and M. Gertler (eds), *The Oxford Handbook of Economic Geography.* Oxford: Oxford University Press. pp. 230–52.

Yeandle, M., Horne, J., Danev, N. and Knapp, A. (2009) *The Global Financial Centre Index: 5.* London: Corporation of the City of London.

6

THE GEOGRAPHY OF UK GOVERNMENT FINANCE: TAX, SPEND AND WHAT LIES IN BETWEEN

Steve Musson

AIMS

- To discover how money is raised to fund spending and investment in the public sectors

- To analyse how and where this money has been spent

- To understand the political tensions and controversies that surround this subject

---

During the financial year 2008–9, United Kingdom government spending exceeded £600 billion; more than £10,000 for every person living in the country. If it were a company listed on the London Stock Exchange, government would have had a larger turnover than the 10 largest corporations combined. At the same time, over 5.7 million people, almost one-fifth of the UK labour force, worked for the government, as teachers, civil servants, doctors and the like. Meanwhile, in the midst of a serious economic downturn, government borrowing had grown to equal almost half of the entire economic output of the country for a year. In short, government is, and has long been, a major part of the UK economy. Although economic geographers have long recognised the importance of government as a regulator of markets, guarantor of social welfare and catalyst for economic development, fewer accounts have emerged that consider the economic geography of government in its own right.

This chapter asks two important questions: where does the government get its money from, and how is it spent? The answers are vital to understanding the UK's economic geography, because so much of what government does has a wider importance. For example, taxes raised in one part of the country are often redistributed elsewhere, establishing financial relationships and interdependencies between different regions and nations (Hamnett 2009) (and see Chapter 3). Government may take direct action, deciding where public investment in hospitals, schools, roads and the like should be made. In a less direct way, the location of public sector employees may affect local labour and housing markets (Marshall et al., 2005). Government intervention in particular industries can also have significant spatial implications. The closure of coal mines in the 1980s and 1990s mainly affected mining areas like Yorkshire, North-east England and South Wales (Bennett *et al.*, 2000), while assistance for banks in the late 2000s implicitly benefited finance centres like the City of London. As such, there is an important role for geographers in understanding the spatial implications of government finance and in relating these patterns to the wider economic geography of the UK.

Tax, borrow and spend: understanding government finance

Let's begin by defining what we mean by 'government'. In the UK, the focus is usually on the two houses of Parliament, the 100 or so members of both houses who hold office as ministers, and the many thousands of civil servants who support the work of government. Collectively, these are usually referred to as 'central government'. Most, but by no means all, central government activities are located in London and they consume almost the entire public sector budget (McLean, 2005). Other types of government exist too, internationally across the European Union and sub-nationally within the UK. For example, devolved forms of government were established in Scotland, Wales, Northern Ireland and London from the late 1990s. Meanwhile, a complex network of district, borough and county councils, collectively known as 'local government', has its own long history of delivering services. Two key features identify these different forms of government. First, their leaders are democratically elected by, and theoretically accountable to, the populations they serve. Second, and crucially for this chapter, they are able to raise their own income, by taxing the people and businesses within their territories.

Many areas of government publish detailed financial information. This fulfils the democratic requirement to be transparent and accountable to the tax-paying electorate and reflects how complex the task of governing the UK has become. The most important source of information is the Budget. This takes the form of a high-profile statement, delivered to the House of Commons each spring by the Chancellor of the Exchequer. The Budget Statement is accompanied by a report, outlining future plans and reviewing the activities of previous years. This includes information on the finances of central, local and devolved government, on predicted sources of income and some data on expenditure. Although the Budget Report usually runs to several hundred pages, it is only the tip of the iceberg in terms of financial information.

Every central government department, devolved administration and local government authority publishes its own detailed plans. This indicates how the money allocated by the Chancellor of the Exchequer will be spent and, in some cases, how additional income will be raised. For example, local government can raise funds through local taxes including Business Rates and Council Tax. At the same time, the devolved administrations in Scotland and London (though not in Wales or Northern Ireland) can raise additional funds through a small levy on Council Taxes in their areas.

Table 6.1 categorises the income and expenditure of central and local government between 2000 and 2009. The left side of the table shows that almost half of all government income came from two sources, Income Tax and National Insurance. Both of these are essentially taxes on the wages of people in employment. Other income sources are individually less important. These include Value Added Tax (or VAT) and Excise Duties on alcohol, tobacco and fuel duty, which can be classified as taxes on consumption. Council Tax and Business Rates are both collected by local government and are calculated according to the value of business and residential property. The funds raised through these local forms of taxation are not nearly enough to meet the needs of local government, which relies heavily on grants from central government to supplement its income. Finally, corporation tax is paid by UK-registered businesses as a proportion of their profits and comprised less than one-tenth of government income.

The right side of Table 6.1 shows combined government spending, broken down into broad functional categories. Social protection, which includes things like Jobseekers Allowance, incapacity benefit, housing benefit and old age pensions, comprised around one-third of total spending. Other forms of welfare, including health and education, were also important. Some high-profile policy areas, including transport, defence and international development, appear

Table 6.1 Public sector balance sheet 2000–9

Income	Per cent of total	Per cent of total	Spending
Income Tax	28	32	Social Protection
National Insurance	17	17	Health
Value Added Tax (VAT)	15	13	Education
Corporation Tax	9	6	Public Order
Excise Duties	9	6	Defence
Council Tax	4	5	Debt Interest
Business Rates	4	4	Housing/Environment
Other	14	4	Economic Development
		3	Transport
		1	International Affairs
		9	Other
Total income in 2000	£371 billion	£370 billion	Total expenditure in 2000
Total income in 2009	£496 billion	£671 billion	Total expenditure in 2009

Source: HM Treasury 2008a

modest by comparison. While almost three-quarters of spending was made directly by central government, local government played a significant role in certain areas like education. However, much of this spending takes the form of 'ring-fenced' grants that can only be used for tightly defined purposes, limiting the amount of control local government has over financial decisions.

We have already seen that taxes on employed people are an important source of revenue for government and that social protection payments are the most significant area of spending. It is worth recognising how closely related the income and expenditure sides of the public sector balance sheet really are. For example, if a person loses their job due to redundancy or ill health, they stop paying Income Tax and National Insurance into the system. There is a good chance that they will have less money to spend, so their VAT contribution (and maybe their fuel, alcohol and tobacco duty, depending on their lifestyle) will also be reduced. At the same time, they may start to take more money out of the system in the form of Jobseekers Allowance, incapacity and housing benefit, or in additional demands on the health system. This isn't to say that people who are ill should be forced to work, or benefit payments denied to those who have the misfortune to lose their jobs. However, from the perspective of the Chancellor of the Exchequer, it is easy to see the financial logic of keeping people in work. Rising unemployment is a financial disaster for government, because it simultaneously reduces income and increases expenditure.

So far, we have only evaluated government finance in terms of the proportions of income and expenditure falling into different categories. By the middle of 2008, government had borrowed to cover its budget deficit and £518 million annual interest payment amounted to 5 per cent of all expenditure. This indicates that in previous years, government had not raised enough income to cover its outgoings. Spending exceeded income for every year from 2000 to 2009 and was expected to grow by a further £175 billion in 2010. In such circumstances, government must choose how to fund the shortfall. One way might be cut back on spending and/or to raise taxes, but both options may be politically unpalatable if they alienate potential voters. A more attractive alternative may be to borrow money to cover the short-term deficit, in the hope of repaying it in the future when lower spending and higher tax revenues create a budget surplus. This can be combined with more economically proactive action, for example by reducing taxes on consumption to trigger growth in consumer spending, or by providing assistance for firms so that they can retain (or even increase) their workforce. For example in 2008, central government intervened to reduce the rate of VAT from 17.5 per cent to 15 per cent and offered assistance for car makers including Vauxhall and Nissan to protect jobs. No government can follow such a strategy indefinitely, because the costs of borrowing to cover additional expenditure and of lost revenue can be significant. There is also a risk that the budget may not return to surplus quickly enough to cover repayments and that tax rises or spending cuts will be even greater as a result.

The geography of government finance

In the first part of this chapter, we considered government income and expenditure for the UK as a whole. For many years, HM Treasury did the same, paying little attention to geographical differences in spending. In some respects, this approach is understandable. After all, decisions on income tax, fuel duty, pensions and the like have traditionally been applied equally in all places. However, since 1990, the Budget Report has included more detailed spatial information, including a regional breakdown of geographically identifiable spending. Some things that can't easily be identified with one part of the UK are still excluded from this analysis, including international development, defence and national security. But for the majority of spending, there are at least two reasons why geography matters. First, national policies frequently have spatially uneven outcomes once they are put into practice. For example, Hamnett (1997) argued that a national reduction in the rate of income tax for those with high incomes in the 1988 Budget mainly benefited people living in London and the South-east. Second, regional differences in government income and expenditure have become increasingly politicised and controversial, particularly since devolved government was established in London, Scotland, Wales and Northern Ireland (McLean and McMillan, 2003).

Table 6.2 explores geographical differences in government spending, using data from 2008–9. Total geographically identifiable spending was £475 billion, meaning that around £143 billion was classified as 'national' and not associated with any one location. England dominated overall spending and received

Table 6.2 Geographically identifiable spending 2008–9

	Total spending (£ billions)	Total spending (per cent UK Total)	Per capita (average in £)	Per capita (per cent UK average)
North–east England	22	5	8,481	108
North–west England	57	12	8,301	106
Yorkshire and the Humber	39	8	7,623	97
East Midlands	30	6	6,983	89
West Midlands	41	9	7,588	97
East of England	37	8	6,617	84
London	69	15	9,176	117
South-east England	54	11	6,555	84
South-west England	36	8	7,026	90
England (all regions)	385	81	7,584	97
Scotland	47	10	9,218	118
Wales	26	5	8,616	110
Northern Ireland	17	4	9,897	126
UK Total	475	100	7,839	100

Note: Approximately £143bn of spending was not geographically identifiable in this financial year.
Source: HM Treasury, 2008a

£385 billion, or 81 per cent of the UK total. Significant regional differences exist within England, where London received around three times as much money as the North-east of England. The majority of spending outside England was in Scotland. Only £17 billion, or 4 per cent of the UK total, went to Northern Ireland. Taking regional population sizes into account, average spending across the whole of the UK is £7,839 per person. In England per capita spending is 97 per cent of the UK average, while in other national territories government spending is significantly higher. Indeed, although Northern Ireland receives a relatively low level of absolute spending, this equates to almost £9,900 per capita, or 126 per cent of the UK average. Similar differences can be observed within the English regions, most notably between London and its neighbours. For example, in the South-east and East of England, average per capita government spending was over £2,600 less than for London in 2008–09.

To understand and explain these geographical differences in government spending, two important financial drivers must be considered. The first is demand for money and/or services. This mainly comes from areas where uptake of state education and healthcare is greatest. As we have seen, these welfare services consume a significant proportion of all government spending, not least because large numbers of public sector workers are employed as teachers, doctors, nurses and so on. Another significant demand is made by social protection, for example unemployment and incapacity benefits. In areas of high unemployment, it follows that social protection spending will be higher and that tax revenues are likely to be lower. In a detailed discussion of the geography of welfare spending, Hamnett (2009: 26) argues that: 'there is a distinct geography to [welfare] spending, both at an aggregate level and in terms of individual benefits and that social security/disability benefits constitute a substantial ... proportion of household income in some less affluent regions.' A second important driver of geographical differences is the system through which money is allocated. Since 1978, HM Treasury has allocated additional money to Scotland, Wales and Northern Ireland though a mechanism known as the Barnett Formula, named after its creator, a Labour politician called Joel Barnett. The purpose of the formula is to provide additional money that compensates for the cost of providing equivalent services to smaller and more sparsely distributed populations, for the disproportionate impact of deindustrialisation and, in the case of Northern Ireland, for the historically higher costs relating to security (Heald, 1994).

Debates around the allocation of funding in the UK often focus on the Barnett Formula, which seems almost universally unpopular (McLean, 2005). In some English regions, it has been suggested that the system, which gives economic advantages to neighbouring Scotland, is unfair. As John Major, Prime Minister from 1990 to 1997, noted in his political memoirs:

> Feeling [about the Barnett Formula] was especially bitter among Members of Parliament with constituencies in English Regions ... which had social problems every bit as serious as Scotland's, but which lost out badly in the

annual public spending rounds, and lacked the economic clout exercised by the government agency Scottish Enterprise. (Major, 2000: 419)

Opposition also exists in those places that the Barnett Formula aims to prioritise. Bell and Christie (2001) note that the formula is increasingly unpopular in Scotland, Wales and Northern Ireland, because it fails to provide sufficient flexibility for devolved national governments to follow their own policies. In this sense, the Barnett Formula can be seen as a mechanism of financial control for central government. Furthermore, McLean (2005) argues that the amount of additional funds provided is gradually being eroded, particularly in Scotland, through a statistical anomaly known as the 'Barnett Squeeze'.

Within England itself, the controversy over funding takes a different form. Amin et al. (2003) suggest that London is central to the 'spatial grammar' of the UK state, dominating the day-to-day political, economic and cultural life of the whole country. This position is reflected in the way London is funded by government. For example, it receives more than twice as much per capita funding than any other English region on transport and more than 50 per cent more for policing and public safety. London also receives more per capita funding for education, training and economic development than any other part of the UK. The formula though which local government is funded works in London's favour too, to compensate inner London local authorities for coping with the needs of large numbers of tourists and other visitors. Furthermore, Marshall et al. (2005) suggest that the location of senior civil service jobs in London brings disproportionate benefits to the capital's labour market. They argue that: 'the concentration of the public sector in London and the South-east contributes to regional imbalance in the British economy with more rapid economic growth close to the capital, which produces simultaneous overheating in these regions and underutilization of infrastructure and human resources elsewhere' (2005: 785). As such, London is privileged by both spatial patterns of government spending and the wider geography of public sector employment.

London contrasts with other regions in southern England, which receive lower levels of per capita spending. For example, south-east England is a prosperous region which borders London to the south and west, extending in a broad arc from Milton Keynes to the Kent coast. Here, local government leaders argue that economically successful regions have low levels of unemployment and relatively affluent populations, so they unfairly contribute more in taxes than they receive in spending. By one calculation, London's net contribution for 2006–7 was £19.7 billion and the South-east's was £14.9 billion (OEF, 2008). In practice, it is difficult to separate out government finance in this way, because there is significant economic overlap between regions. For example, companies based in London and the South-east may depend on skilled workers, suppliers and customers elsewhere in the UK for their success. However, given the competing claims of different nations and regions of the UK, it is perhaps no wonder that politicians in every region think the current system is in some way unfair. As McLean (2005: 225) notes: 'No player is satisfied with the result. The three [national] Territories all dislike Barnett, albeit for opposite reasons (Scotland because the 'squeeze' is too harsh). The English regions all think they get a raw deal and blame it on Barnett.'

Investing in public services: capital spending and public–private partnerships

The ways in which government collects and spends money are insightful, because they offer a tangible way of measuring policy decisions. It is one thing for a minister to promise more money for education or health, but another to see how this works in practice. However, as we have already seen, some areas of government finance offer more potential for discretion than others, depending on how demand-led they are. HM Treasury makes a useful distinction in this respect, dividing all government spending into two types, as either current or capital spending. Current spending is by far the greater of the two, accounting for around 90 per cent of total government finance in 2008–9. It includes most of the day-to-day running costs of central and local government, such as staff salaries, maintenance charges and purchasing small items of equipment. Current spending also includes social protection payments, which are usually paid on demand according to entitlement. As such, the rules for claiming Jobseekers Allowance do not vary from place to place, even if the numbers of claimants does, depending on local economic circumstances.

Capital spending is different. Although it accounts for only 10 per cent of total spending, it offers a different set of insights into the ways that government operates. Capital spending includes major public investment, for example in new hospitals and schools, as well as national projects like military hardware. Given the enormous costs often involved, it is not surprising that capital spending is dominated by central rather than local government. Investment decisions can often be highly politicised and controversial, especially if contracts are awarded to suppliers from outside the UK, or if investment in one location means that another area misses out. In one high-profile example from 2000, a £550m nuclear research facility was developed in Oxfordshire in south-east England rather than Cheshire in north-west England, which regional politicians and trade unionists in the North-west saw as confirmation of southern bias in capital investment decisions (Perry, 2007). The crucial thing to emphasise is that, unlike current spending, which is dominated by demand for and entitlement to services and money, capital spending offers more room for political discretion. A crisis like a fire or flood might require urgent investment in a particular location. However in most cases, a decision has to be made – frequently at ministerial level or above – about where to build a new hospital or to locate a new research facility.

Table 6.3 presents data on capital spending in 2008–9. It shows that London receives a higher proportion than anywhere else in England, while Scotland

Table 6.3 Geographically identifiable capital spending 2008–9

	Capital spending (£ billions)	Capital spending (per cent UK Total)	Per capita (average in £)	Per capita (per cent UK average)
North-east England	1,860	4	739	88
North-west England	5,233	11	777	92
Yorkshire and the Humber	3,186	6	641	76
East Midlands	2,491	5	597	71
West Midlands	3,381	7	642	76
East of England	3,061	6	567	67
London	8,307	17	1,156	137
South-east England	4,590	9	573	68
South-west England	2,864	6	580	69
England (all regions)	34,973	71	711	84
Scotland	5,572	11	1,101	131
Wales	2,193	4	755	90
Northern Ireland	1,609	3	955	113
UK Total	49,553	100	842	100

Note: Approximately £5.2bn, or 11 per cent of total capital spending, was not geographically identifiable.
Source: HM Treasury, 2008a

and Northern Ireland also receive significantly more than the UK average on a per capita basis. Such is the concentration of government investment in these areas that no other part of the UK receives more than the national average. In spite of concerns about southern bias noted above, the most under-funded areas shown in Table 6.3 are the East of England, the South-east and South-west, which constitute the rest of southern England outside London. Another significant loser in terms of per capita spending is Wales. Although its share of total government spending is protected by the Barnett Formula, in capital spending terms it receives only 90 per cent of the UK average. For the rest of England, the further away from London a region is, the better it fares. As such, the East and West Midlands receive more capital spending per capita than the regions of southern England, but less than the North-east and North-west. The concerns of northern English regions, that they receive less funding than neighbouring Scotland, are brought into sharp focus when more discretionary elements like capital spending are analysed.

In keeping with the rest of the chapter, the discussion of capital spending has so far focused exclusively on government finance. It is worth considering the growing importance of the private sector in this respect, especially since the introduction of the Private Finance Initiative (or PFI) in 1992. PFI is a form of public–private partnership, where private contractors bid to design, build and in some cases operate a public sector asset, like a new school, hospital or road. The private sector raises funds for investment and recoups its outlay by leasing the development back to central or local government over a long period, often 25–30 years. The terms of services to be provided and the timetable of payments are usually tightly defined by a contract. In the context of government finance, the short-term attraction of this arrangement is clear. Money does not have to be raised from sources like income tax, but is instead provided by the private partner. But from the outset, public–private partnership proved to be controversial (Pollock, 2004). Academics and trade unionists argued that PFI was more expensive in the long run, because private companies would need to make a profit on their investment. Meanwhile, the lengthy PFI contracts were inflexible to changing demands for services. More fundamentally, it was suggested that the financial logic of private enterprise was incompatible with the social principles of public service. In this sense, public–private partnership was seen as a 'selling out' of the welfare state, because it put decision making into the hands of private investors.

Notwithstanding these concerns, PFI and other forms of public–private partnership became increasingly commonplace under the Blair–Brown Labour administration and, between 1992–2008, almost £60 billion of additional public investment was secured (HM Treasury, 2008b). In keeping with other forms of capital spending, investment was heavily focused in London. By the end of 2008, 42 per cent of all PFI was located in the capital. The renewal of London Underground was particularly important, involving the three largest PFI contracts ever signed with a combined investment value of £17.6 billion. The cumulative value of PFI in other parts of the UK is less than in London, and the spatial pattern of projects much patchier. However, in those places where PFI investment has taken place, the local impacts are far-reaching. For example, a £333 million contract signed in 2003 funded a new hospital in Derby with 1,150 beds and 35 operating theatres. The significance of this investment goes beyond any effects it might have on the quality of medical treatment, because hospital care in Derby is now provided by a private consortium under the terms of the contract signed with central government. While it does not follow that the quality of service is necessarily better or worse than under traditional public ownership, there are nevertheless some important implications. For example, near the beginning of this chapter, we noted the link between government spending and accountability to taxpayers, through the publication of transparent financial information. However, PFI contracts are commercially confidential and few details on service standards and costs are publicly available. As such, under public–private partnerships, the geography of investment in public services is about more than just money; it is also about accountability and transparency.

Conclusion

This chapter began by emphasising the importance of government as an economic actor in its own right. We noted spatial differences in the way government raises and spends money, which reflect the geography of demand for and entitlement to services and money. We also considered more discretionary forms of government spending, for example investment through both capital spending and public–private partnerships. In both cases, London receives more per capita spending than any other area, reflecting Amin et al.'s, (2003) understanding of London as being central to the 'spatial grammar' of the UK state. We also considered why the geography of government finance matters, for example by determining who provides public services in different locations, who pays for them, and how accountable and transparent such arrangements are. Government does more than simply set rules and regulations for the economy and pick up the pieces when things go wrong; it is an active player in the economic geography of the UK. Ongoing research, for example in the role of the public and private sector in education and healthcare and in geographical differences in the nature of the welfare state, continues to explore these issues.

Further reading

There is a large literature on the ways that government collects and spends money, some of which is very complex. These suggestions for further reading deal with some of the big issues, like welfare spending, devolution and the Barnett Formula and the centrality of London, in an accessible way.

- Hamnett (2009) offers an up-to-date analysis of the geography of welfare and housing benefit expenditure in the UK.
- Marshall et al. (2005) review the arguments surrounding civil service decentralisation from London.
- McLean (2005) assesses the challenges facing the UK's system for allocating public expenditure (see especially Chapters 1 and 5).

References

Amin, A., Massey, D. and Thrift, N. (2003) *Decentering the Nation: A Radical Approach to Regional Inequality*. Catalyst Paper 19, September. Available from: The Catalyst Forum, 150 The Broadway, London SW19 1RX.

Bell, D. and Christie, A. (2001) The Barnett Formula: nobody's child? in A. Trench (ed.), *The State of the Nations 2001: The Second Year of Devolution in the United Kingdom*. Thorverton: Imprint Academic.

Bennett, K., Hudson, R. and Beynon, H. (2000) *Coalfields Regeneration: Dealing with the Consequences of Industrial Decline*. Bristol: Policy Press and Joseph Rowntree Foundation.

Hamnett, C. (1997) A stroke of the Chancellor's pen: the social and regional impact of the Conservatives' 1988 higher rate tax cuts, *Environment and Planning A*, 29: 129–47.

Hamnett, C. (2009) Spatial divisions of welfare: the geography of welfare benefit expenditure and housing benefit in the United Kingdom, *Regional Studies*, 43: 1015–33.

Heald, D. (1994) Territorial public expenditure in the United Kingdom, *Public Administration*, 72: 147–75.

HM Treasury (2008a) *Budget 2008 – Stability and Opportunity: Building a Strong and Sustainable Future*. HC388, 12 March. London: The Stationery Office.

HM Treasury (2008b) *PFI Signed Projects List – November 2008*. Accessed 20 May 2009 at www.hm-treasury.gov.uk/ppp_pfi_stats.htm

Major, J. (2000): *John Major: The Autobiography*. London: HarperCollins.

Marshall, J., Bradley, D., Hodgson, C., Alderman, N. and Richardson, R. (2005) Relocation, relocation, relocation: assessing the case for public sector dispersal, *Regional Studies*, 39: 767–87.

McLean, I. (2005) *The Fiscal Crisis of the United Kingdom*. Basingstoke: Palgrave Macmillan.

McLean, I. and McMillan, A. (2003) The distribution of public expenditure across the UK regions, *Fiscal Studies*, 24: 45–71.

OEF (Oxford Economics Ltd) (2008) *Regional Winners and Losers in UK Public Finances*. Special Report, 20 July. Accessed 20 May 2009 at www.oef.com/free/pdfs/ukmpubfinfeat(jul).pdf.

Perry, B. (2007) The multi-level governance of science policy in England, *Regional Studies*, 41: 1051–67.

Pollock, A. (2004) NHS PLC: *The Privatisation of our Healthcare*. London: Verso.

7

STATE AND ECONOMY: GOVERNING UNEVEN DEVELOPMENT IN THE UK

Andy Pike and John Tomaney

AIMS

- To establish the relations between state and economy as central to understanding how geographically uneven development is governed

- To demonstrate how the spatial disparities in economic and social conditions inherent in capitalist economies reproduce intractable problems for states in governing their territories

- To show how the UK state has unleashed devolution and new forms of spatial economic policy in its recent attempts to manage the geographical tensions generated by uneven development

In 1999, during a period of relatively strong economic growth, the then Prime Minister Tony Blair visited Manchester as he sought to dispel stereotypical views of the North–South divide in economic and social conditions in the UK. The Prime Minister argued that the North–South divide was 'an over-simplistic explanation of the problems that regional economies face' and that 'The real divide is between the haves and the have-nots, whatever part of the country you are in' (BBC News, 6 December 1999). Backed up by a Cabinet Office report that attempted to demonstrate that 'the disparity within regions is at least as great as that between them', Tony Blair's aspiration for the New Labour Government was 'to make sure that all parts of Britain share in rising prosperity and in all parts of Britain we tackle poverty' and ensure that 'We are narrowing the gap for the very poorest of people in the country'.

The geographical depth and scope of disparities in economic and social conditions are of long-standing importance in the UK (as also discussed in Chapters 2 and 3). Since the Great Depression of the 1930s reshaped the landscape of prosperity from North to South, the evolution of spatially uneven development has mattered to government. As Tony Blair's comments in the late 1990s illustrate, disparities between people and places in prosperity and wellbeing are economically, socially and politically resonant. To understand how and in what ways such geographical disparities are governed in the UK and elsewhere, first, we establish the central importance of the relationship between the state and the economy. Second, we show how the economic and social disparities inherent in capitalism reproduce intractable problems for states in governing their territories. Third, we examine how the UK state has unleashed devolution and new forms of spatial economic policy in its recent attempts to manage the geographical tensions generated by uneven development. We use the idea of the 'qualitative state' to examine the UK state in attempting to govern uneven development. This approach analyses state strategies, ideologies, guiding theories, development aspirations and political projects as well as organisation, structure, capacities and processes. A consideration of historical context reveals the evolution of state-economy relationships and changing eras in governing uneven development in the post-war period; from Keynesianism and Thatcherism within a centralised state to the recent emergence of devolution and 'new regional policy'. We then examine the complexities and tensions involved in governing uneven development in England within the UK state before concluding and reflecting on potential future developments.

State and economy – governing uneven development

Historically, the state and economy relationship central to understanding how uneven development is governed has been thought about in terms of the quantitative extent to which the state intervened (or not) in the workings of the economy. The relationship was typically represented in models of *political-economy* such as centrally planned economies like those of state socialism in the former eastern bloc or mixed economies like those of western Europe or North America. In contrast to thinking about degrees of intervention, institutionally-oriented approaches see states as fundamentally implicated in capitalist economy, for example through guaranteeing private property rights, regulating markets, controlling territorial boundaries, providing basic infrastructure and managing macro-economic policy. This more qualitative approach emphasises the historical changes and evolutions in the nature, purpose and consequences of the forms of state agency. It focuses on what the state does and how it does it, for example its strategies and political projects, organisation and structure, capacities and processes.

Situating the 'qualitative state' in geographical context demonstrates how spatial disparities confront states with intractable problems (O'Neill, 1997).

Capitalism is reproduced and regulated through circuits of production, exchange and consumption driven by the accumulation imperative and disruptive forces of competition and innovation. The constant remaking and inherent geographical unevenness in the socio-spatial organisation of economy and society under capitalism makes it a recurrent concern for states (Harvey, 2001). Uneven development – that is, spatial disparities in economic and social conditions including prosperity and health – is manifest at a range of scales (for example regional, local and urban) and, depending upon its extent and nature, constitutes a central issue for capitalist nation states in governing their territories. Why? The spatial distribution of economic activity shapes the character, pattern and trajectory of national, regional and local growth, development and prosperity. Politically, marked geographical differentiation in wellbeing, life chances and access to public services represent inequalities between people and places and raise questions of territorial justice. In particular circumstances, then, states can recognise and direct resources to address spatial disparities through forms of policy either directly spatial – such as regional and urban policy – or indirectly spatial – such as welfare and defence. As we shall see in our discussion of the UK state wrestling with uneven development, the diagnoses of geographical inequalities and policy are highly differentiated and spatially variable over time and space (and see also Chapter 6).

'Hollowing out' has been very influential in explaining changes in the geographies of state-economy relations governing uneven development (Jessop, 2002). In the context of globalisation, the nation state is seen as losing or actively delegating powers 'upwards' to supranational bodies (e.g. the World Trade Organization, the European Union), 'downwards' to sub-national institutions (e.g. regional, city and local governments) and 'outwards' to cross-boundary and/or transnational organisations (e.g. cross-border co-operation networks). 'Hollowing out' has been accompanied by a shift from government – characterised by state institutions with direct relations to their citizens – to governance – where states steer arms-length organisations with indirect connections to their populations. State boundaries have blurred as functions and services are increasingly delegated to more specialised quasi-state or non-state organisations or privatised and/or outsourced to the private and voluntary sectors. 'Hollowing out' is not simply the emptying of the nation state, however, with governance creating a more complicated picture of 'filling-in' in geographically differentiated ways and at new scales such as the supranational, regional, local and neighbourhood (Goodwin et al., 2005). In territorial terms, such changes represent the 're-scaling' of the state and the emergence of 'new state spatialities' at different geographical levels including the region and the city-region (Lobao et al., 2009). Transcending the notion of bounded, fixed territories, other work interprets the emergence of unbounded, fluid 'spatial assemblages' of more complex and interconnected networks of governance (Allen and Cochrane, 2007). In sum, the 'qualitative state' sees state-economy relations facing the intractable problem of governing the inherent uneven development under capitalism because states are inescapably entangled in the economy. In institutional terms, the 'qualitative state' is multi-level across and between spatial scales and multi-agent involving state, para-state and non-state institutions.

Keynesianism and Thatcherism in a centralised state

The UK is a multi-national 'union state' with highly centralised structures and traditions, differentiated by separate territorial administrations established in Scotland, Wales and Northern Ireland in the late nineteenth century (Tomaney, 2000). Such history has engendered strong belief in the central national state and its departmental organisation in the design and delivery of public policy. Although regional policy first emerged in the inter-war period to address unemployment 'blackspots' caused by localised rationalisation of traditional industries, we focus our account on its rise and fall after 1945. Governing uneven development in the post-war period was characterised by 'spatial Keynesianism' (Martin and Sunley, 1997) focused upon investment and the stimulation of cumulative causation to manage aggregate demand and employment in the regions (and see also Chapter 3). The geographical unevenness of what the Royal Commission on the Distribution of the Industrial Population in the early 1940s called 'over-development' in the South and 'under-development' in the North were interpreted as inextricably connected. Spatially unbalanced and cumulative growth drew economic resources from peripheral to core regions. Keynesian spatial economic policy relied upon the contained territory of the national economy within which investment and growth could be redistributed from growing, richer to lagging, poorer regions to alleviate spatial disparities (Figure 7.1). The initial case for regional policy

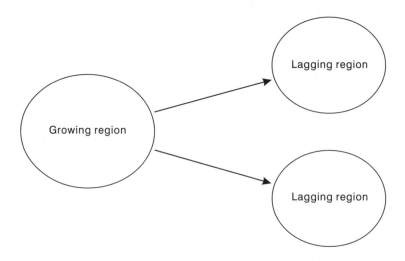

Figure 7.1   Redistributive spatial economic policy (Adapted from Pike et al., 2006)

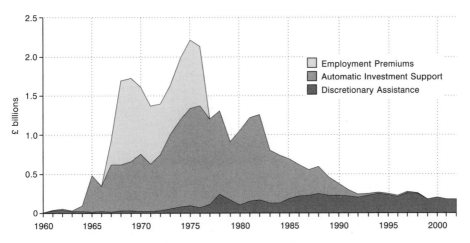

Figure 7.2 Expenditure on Regional Industrial Assistance, 1960–2002 (Wren, 2005, reprinted with permission. Figures are for actual grant payments at constant prices for Great Britain.)

was economic: correcting the inefficiencies of under-utilised resources and lack of economic modernisation. The later social case evolved to ensure economic equity between regions and ameliorate the hardship generated by geographically localised deindustrialisation and job loss in the traditional industrial heartlands of northern England, south Wales and Scotland (Morgan, 2006). Using the 'carrots' and 'sticks' of investment and labour subsidies and controls, spatial economic policy had a regional focus on different tiers of Assisted Areas and was directed 'top-down' by national civil servants in central government. Bound by the 'One Nation' politics of the post-war settlement, this era of classic regional policy was marked by periods of fluctuating emphasis and expenditure under both Conservative and Labour administrations prior to its high water mark in the mid-1970s (Figure 7.2). Redistributive regional policy was accompanied by limited institutional change beyond the short-lived Regional Planning Councils operating during the 1960s and 1970s.

Despite some success in job creation and economic diversification in the Assisted Areas, the political-economic tide turned away from Keynesianism and redistributive regional policy. Neo-liberalism emerged following the crisis of stagflation, industrial strife and public fiscal imbalances during the 1970s. Characterised by de-regulation, liberalisation and the attempted 'rolling-back' of the state, the UK variant was led by Margaret Thatcher's Conservative administrations from 1979. Principles of individual responsibility, free markets and enterprise underpinned the critique and dismantling of regional policy during the 1980s (Martin, 1988). The turn toward neo-classical economics and the free-market interpreted Keynesianism and regional policy as distortions and impediments to rational and efficient decision-making amongst economic actors. Subsidies were seen as economically inefficient and wasteful, causing 'dead-weight' effects in supporting

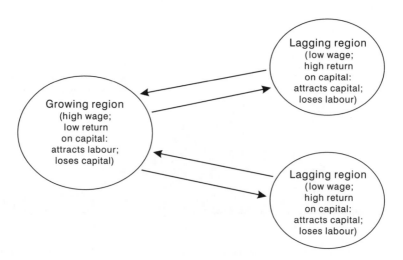

Figure 7.3 Free-market spatial economic policy (Authors' research)

activities that would have occurred anyway and unable to tackle structural problems including lack of enterprise and innovation.

In contrast to the Keynesian emphasis upon demand, neo-classical theory emphasised the flexibility and upgrading of the supply side of factor markets such as labour skills. Policy sought to promote the spatial mobility of factors of production in rational response to price signals. Convergence in regional economic conditions would occur through the adjustment mechanism of the movement of capital and labour in opposite directions between high and low wage regions (Figure 7.3). From the 1970s, structural change toward services in the UK economy favoured regions and localities around London and the Greater South East, sharpening a North–South divide in regional disparities (Martin, 1988). Spatial economic policy under Thatcherism reduced spending, shrunk the map of Assisted Areas and changed support from automatic to discretionary selectively to encourage small enterprise and the attraction of international inward investment in the Single European Market (see Figure 7.2). The geographical focus of spatial policy moved to the urban scale and the institutional lead and resources were redirected from local government to new special-purpose bodies such as Urban Development Corporations and Local Enterprise Agencies. Formerly state-owned industries such as coal, steel and shipbuilding were privatised and rationalised with highly damaging localised impacts (Hudson, 1989). Alongside its liberalising ethos, Thatcherism still relied upon the highly centralised apparatus of the UK state for tight control over spatial policy and local government. Thatcherism recast state-economy relations and its spatial economic policy served to widen regional disparities in the UK during the 1980s. John Major's successor Conservative administrations continued working within the national central government framework with an urban focus, introduced more competition for resources and began the regionalisation within England with the establishment of Government Offices for the Regions in 1994.

Devolution and 'new regional policy'

After New Labour's election victory in 1997, devolution and constitutional change constituted part of its state modernisation project. It unfolded as an 'asymmetric' or differentiated set of institutions with varying powers and resources comprising the Parliament in Scotland, National Assemblies in Wales and Northern Ireland and, reflecting its exceptional size and status, the Mayoralty and Elected Assembly in London. This spatially uneven political geography created a multi-level, polycentric UK state. Working across several geographical scales, it represented a more distributed landscape of political power (Morgan, 2007). Although identity was integral to claims for self-determination in Scotland and Wales, the UK largely mirrored the international shift toward economy as the primary rationale for the global decentralisation trend during the 1990s (Rodríguez-Pose and Sandall, 2008). A potential 'economic dividend' was attached to devolution. Delivered, first, through enhanced autonomy to design and deliver policy better tailored to regional and local needs. And, second, through more responsive, effective and accountable governance systems providing decentralised institutional capacity to mobilise and shape collective action for development. For the peripheral and poorer nations and regions of the UK, devolution held out the prospect of tackling the persistent spatial disparities entrenched following the early 1990s recession as London and the Greater South-east recovered more strongly and led the sustained growth from the late 1990s until the crisis and recession of 2007–10.

In tandem with devolution, the theoretical basis of governing the economy and effecting spatial economic policy moved from the Thatcherite approach heavily influenced by neo-classical economics towards new endogenous economic growth theory. This 'new economic geography' incorporates technological innovation (rather than treating it as exogenous or external) and emphasises increasing (rather than diminishing) returns in the utilisation of factors of production, the improvement of productivity, agglomeration and spill-over effects in propelling economic growth (Table 7.1). In contrast to the redistribution of Keynesian spatial economic policy but complementing the supply-side focus of neo-classical policy, endogenous growth theory emphasises productivity (as a proxy for competitiveness). Its spatial economic policy is oriented towards the encouragement and support of territories deemed to have growth potential. Rather than attempting spatially to redistribute growth in a more interdependent and internationalised economy, spatial economic policy seeks to enable cities, regions and localities to fulfil their economic potential by raising their own competitiveness and performance (Figure 7.4). It shares the neo-classical preoccupation with correcting market failure, however. In terms of the politics of the qualitative state, Kevin Morgan (2006: 26) sees

Table 7.1　Redistributive, free-market and growth-oriented spatial economic policy

Characteristic	Redistributive	Free-market	Growth-oriented
Economic theory	Keynesian growth theory	Neo-classical (exogenous) growth theory	New (endogenous) growth theory
Causal explanation of spatial disparities	Low aggregate demand and investment, structural weaknesses	Inherited factor endowments and quality, inflexibility and immobility in factor markets	Constructed factor endowments and increasing returns generating productivity and innovation differentials
Adjustment process	Spatial disparities persist through cumulative causation, multiplier, spread and backwash effects	Factor market adjustment returns to equilibrium and convergence reduces spatial disparities	Agglomeration and spill-over effects, national growth and spatial disparity trade-off
Policy rationales	Redistribution for economic efficiency and spatial and social equity	Improving factor market efficiency, flexibility and mobility	Market failures or equity
Policy instruments	Automatic capital and labour subsidies, industrial development controls, infrastructure investment	Regional Selective Assistance, enterprise grants for SMEs and new start-ups	Innovation grants, Venture capital funds
Institutional organisation	Centralised, national	Centralised, national	Decentralised, sub-national, regional, city (-regional) and/or local
Geographical focus and scope	Regional	Regional, local and urban	City(-regional)
Political-economic project	Social Democratic	New Right, Neo-liberal	Third Way, Neo-liberal
Language	Regional inequalities, redistribution	Regional and local divides, trickle-down	Spatial disparities, performance gaps spillovers

Source: Authors' research

this shift as HM Treasury capturing regional policy and recasting it from 'a Keynesian welfare policy into a Schumpeterian development tool to enhance the UK's low productivity levels.' Rather than the 'top-down' centralism of previous 'old regional policy', the 'new regional policy' is 'bottom-up'. New devolved institutions are entrusted with leading and enhancing the productivity of their regions and localities to deliver their own growth and prosperity and to contribute to the national economy. The 'economic dividend' of devolution is the means of redressing spatial disparities through a 'levelling-up' not 'levelling-down' process.

Critiques of devolution highlight the difficulty of discerning any 'economic dividend' (Rodríguez-Pose and Gill 2005) and its alignment with neo-liberal 'new regionalist' state restructuring, decentralising responsibilities without concomitant resources, and promotion of inter-territorial competition (Lovering, 2001).

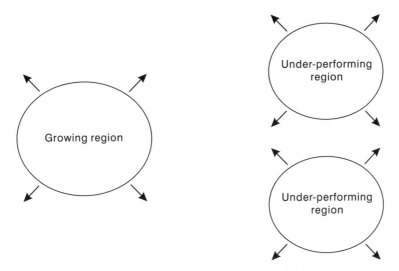

Figure 7.4   Growth-oriented spatial economic policy (Adapted from Pike et al., 2006)

'New regional policy' is interpreted as failing to address spatial disparities because of its narrow focus upon economic growth and productivity in each and every region (Fothergill, 2005). It underplays the importance of the demand side and the number of firms and jobs in places and neglects the importance of spatial industrial structure and divisions of labour in the geographical distribution of types of functions, jobs and occupations. Moreover, it has only a simplistic conception of the public sector's role in spatial development. The experience of England is instructive in understanding how devolution and the 'new regional policy' have fared for the UK state in governing uneven development.

Addressing the 'The English Question'

England's size and weight within the UK – constituting some 80 per cent of the total population and nearly 90 per cent of GDP – renders considerations of governing its uneven development and growth important and difficult for the UK state. John Major's Conservative government signalled at least some recognition of the limitations of centralisation by establishing Government Offices for the Regions (GORs) as representatives of their national government departments and listening posts in 1994. New territorial boundaries redrew the Standard Regions map of England, separating out London, the South-east and eastern regions (Figure 7.5). GORs represented the beginnings of 'regionalisation' – 'rescaling' the state at the regional level – as a putative solution to governing territorial differences and fostering heightened efficiency and enhanced co-ordination between state and para-state bodies.

The complexities and difficulties of resolving 'the English Question' meant that England received barely a mention in the devolution legislation. Instead, regionalisation accelerated markedly only following New Labour's election in

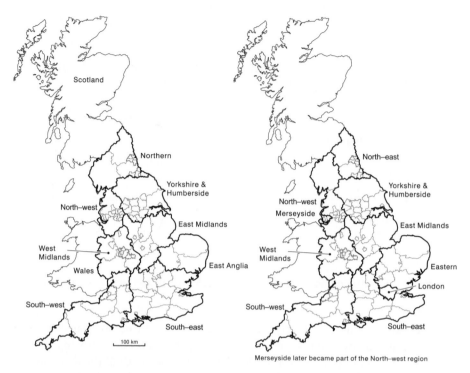

Figure 7.5   Standard Regions (from 1974) and Government Office Regions (from 1994) in England (Redrawn from Hogwood, 1996: 13, 20)

1997. In the new devolved architecture of the UK state, Regional Development Agencies (RDAs) were established as strategic economic leaders in *every* English region – including London under its Mayoralty and Assembly. Indirectly elected Regional Chambers (later rebranded Assemblies) were given some responsibilities for scrutinising RDAs and land-use planning. Despite the Government's commitment to improve growth in all regions *and* reduce gaps in the growth rates between them, concerns about spatial disparities in England persisted. The hierarchy of 'regional competitiveness' and 'under-performing' regions remained little changed during the 1990s and 2000s. In terms of gross value added (GVA) per head – the value of goods and services produced by employees – London and the south-east actually pulled further away from the rest (Table 7.2). Discontent focused upon the inability to ameliorate spatial disparities by 'treating unequals equally' through endowing every region with a growth-oriented RDA (House of Commons, 2003), albeit with modest powers and huge responsibilities but control of less than 1 per cent of public expenditure within their regions. Questions were asked too about encouraging all regions to attempt the mathematically impossible task of growing at a *faster* rate than the national average. Despite HM Treasury's 'devolving decision-making' agenda, the burgeoning tier of regional institutions continued to animate debate about the regional accountability of 'non-departmental public bodies'.

Table 7.2 Gross value added (workplace basis) per head by region and nation, 1989–2006 (GVA Index UK = 100)*

	United Kingdom	North-east	North-west	Yorkshire and the Humber	East Midlands	West Midlands	East of England	London	South-east	South-west	England	Wales	Scotland	Northern Ireland
1989	100	84.5	91.9	91.0	96.5	93.1	96.4	149.7	100.5	92.5	102.0	85.4	97.1	74.6
1990	100	84.1	91.7	90.6	95.9	93.0	97.0	148.7	100.8	93.0	101.9	84.9	98.0	75.5
1991	100	84.9	91.3	90.6	95.4	92.3	95.9	149.3	101.3	92.6	101.8	83.9	99.3	76.8
1992	100	85.2	91.4	90.1	94.9	92.2	96.0	149.2	101.8	92.0	101.7	83.9	99.8	77.6
1993	100	84.5	91.1	89.3	94.3	92.3	96.1	149.6	102.5	92.1	101.7	83.5	99.6	78.9
1994	100	84.1	91.2	89.1	94.4	93.1	96.2	148.3	102.9	91.9	101.7	83.7	99.8	79.2
1995	100	83.8	91.0	89.8	94.7	93.6	96.4	146.3	103.3	92.1	101.6	84.0	99.7	80.6
1996	100	82.6	90.3	89.9	94.7	93.2	96.2	147.2	103.9	93.2	101.8	82.7	98.9	81.1
1997	100	81.4	89.7	89.7	94.3	93.2	95.6	149.1	104.7	93.1	102.0	81.3	97.9	80.6
1998	100	80.3	89.1	88.9	93.2	92.6	95.5	151.7	106.5	92.8	102.3	79.5	96.2	80.8
1999	100	79.5	89.1	88.1	92.0	92.0	95.1	153.0	107.7	92.8	102.4	78.5	95.2	81.1
2000	100	79.3	88.9	87.9	91.5	91.8	95.0	152.6	108.7	92.8	102.5	78.3	94.7	81.5
2001	100	79.4	88.9	87.8	91.8	91.5	95.2	151.2	109.5	93.3	102.5	78.4	94.1	81.2
2002	100	79.0	88.4	87.7	91.5	90.5	95.1	153.0	109.3	93.4	102.5	78.1	94.3	80.7
2003	100	79.3	87.9	87.4	91.7	89.6	95.4	154.0	109.2	93.6	102.5	77.8	94.4	80.7
2004	100	80.0	87.7	86.9	91.6	89.1	95.7	154.6	109.1	93.9	102.5	77.5	94.3	81.2
2005	100	80.9	87.3	86.2	91.5	88.9	95.2	155.2	109.0	93.6	102.5	77.3	95.0	81.4
2006	100	81.5	87.1	85.7	91.1	89.0	94.7	155.4	109.0	93.8	102.4	77.3	95.5	81.5

Notes: * The regional GVA series for this table have been calculated using a five-period moving average. Estimates of workplace based GVA allocate income to the region in which commuters work. Regional GVA figures from 1989 to 2005 have been revised due to revisions to national controls (Blue Book 2006) and survey results.

Source: Regional Accounts, Office for National Statistics

Attempts better to co-ordinate or 'join-up' public policy amongst the tangled web of bodies at national, regional and local levels were complex – as the governance map of North-east England demonstrates (Figure 7.6).

Faced with governing the economic and democratic deficits in the English regions, the government proposed strengthened GORs and the creation of Elected Regional Assemblies (ERA). ERAs were afforded limited powers for economic development, spatial planning and transport and modest finance-raising capacities conditional upon the demonstration of sufficient public support and single-tier (unitary) local government. The ERA package on offer was overwhelmingly rejected by a 4:1 margin in the sole referendum in North-east England in 2004 (Figure 7.7). The result reflected its weak powers, uneven enthusiasm and hostility in central government, the lack of faith in national government late in its second term and the broader currents of distrust in politicians and political institutions.

Amidst calls for a radical decentralisation of the state and the spatial dispersal of government Ministries (Amin et al., 2003), faltering regionalisation and regionalism in England created a vacuum into which an array of emergent 'spatial imaginaries' flowed (Pike and Tomaney, 2009). Each claimed to offer an institutional fix for the persistent problem of governing spatially uneven development: (1) resurgent cities and/or city-regions as motors of their regional economies; (2) localism led by local authorities capable of decentralised approaches to economic development; and (3) pan-regionalisms focused upon cross-regional issues including housing, jobs and infrastructure in new 'growth areas' in the Greater South-east, housing market renewal areas in northern cities and the Northern Way seeking to close the productivity gap between the northern regions and London and the Greater South-east.

Faced with such complexity and intractable spatial disparities, in 2007 the Government sought to streamline the governance of economic development in England through its Sub-National Review. Regional tier responsibilities and roles were reshaped with RDAs assuming a more strategic role and delegating to sub-regions as well as individual and groups of local authorities. Meanwhile, the opposition Conservative policy maintained its hostility to regionalism and preference for local government but without a clear and coherent vision for how England and its spatial disparities might be governed. While government grappled with multi-level and multi-agent co-ordination and working between and across institutions working at different spatial levels, the recurrent issue of governing the highly unequal growth within England still confronts the UK state. Against the background of proposed reforms, London's economic and population growth continues to afford it a unique position and weight in explaining and framing analysis and policy for spatial disparities in England and the UK. This has meant that the *regional* pressures of growth in London and the Greater South-east are translated into demands for *national* public investment in infrastructure, especially housing and transport, to enable such growth for the greater good of the UK economy. In an array of large-scale projects, historically unprecedented levels of public investment totalling nearly £100bn have been or are being directed into London and the Greater South-east (Figure 7.8).

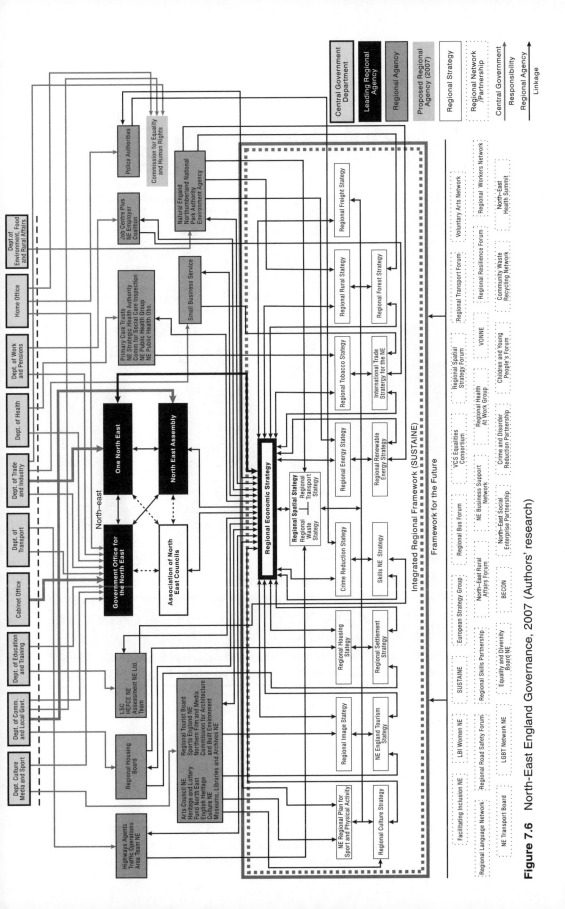

Figure 7.6 North-East England Governance, 2007 (Authors' research)

Figure 7.7   *The Journal*, 5 November 2004 (© Newcastle Chronicle and Journal, 2004, reprinted with permission)

Conclusions

The UK's state's experience of governing uneven development demonstrates the importance of understanding the evolution of state-economy relations and the recurrent difficulty of confronting the spatial disparities inherent in capitalist forms of economy. Conceptualising the qualitative state in a geographically

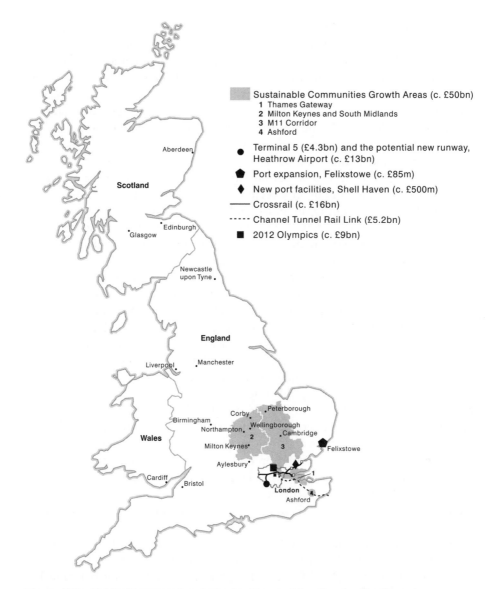

Figure 7.8 Public investment projects, London and the Greater South-east (Authors' research)

sensitive way emphasised the importance of reading the historical evolution of stage agency; its strategies, underlying theories and rationales, political projects, structural organisation, capacities and processes. In the context of the UK's highly centralised and multi-national state, Keynesian regional policy sought to heighten economic efficiency and social equity across territories through redistribution to manage demand and employment. In a marked turning point, Thatcherism maintained a strong centralised state but sought to roll-back its reach in selective areas through de-regulation, liberalisation and privatisation. Drawing upon neo-classical economics, the state moved in the direction of free-market spatial economic policy to improve supply-side flexibility and mobility in factor markets and effectively dismantled the existing Keynesian policy and institutional framework. In the most recent era, the UK state has introduced asymmetric devolution in its nations and regions and constructed a growth-oriented 'new regional policy' based upon new (endogenous) economic growth theory led by nationally-enabled regional institutions. By 2004, regionalisation and regionalism stalled as a means of governing the economic and democratic deficits in England. Since then the UK state has effectively encouraged a bottom-up outpouring of competing 'spatial imaginaries' – cities and/or city-regions, localisms and pan-regionalisms – facilitated to try to identify, mobilise and valorise their economic potentials to contribute to national growth. The current conjuncture is marked by complexity, experimentation, fragmentation and incoherence with largely negative implications for territorial equity and justice.

The qualitative state provides a supple approach to understand the diverse and varied ways in which state-economy relations evolve over time and through space in wrestling with the conundrum of governing uneven development. Conceptually, such a nuanced approach is central to future research agendas. First, explaining how general processes – such as devolution, 'hollowing out' and 'filling-in' the state – are mediated in particular geographical contexts. Devolution, for example, comes in different shapes and sizes; it is a process (not an event) that unfolds unevenly and its outcomes – both positive and negative – are difficult to discern and highly conditional upon specific and particular factors. Recognising 'varieties' of devolution (Cooke and Clifton, 2005) in the nations and regions of the UK goes some way in capturing the different kinds of devolution and their uneven abilities to tackle the powerful forces reproducing uneven development.

Second, future research will be animated by the territorial politics of devolution within the UK. Policy divergences from the 'national' policy frameworks enacted in England – social care and university tuition fees, for example – reflect the politics in the devolved territories within and between the UK-wide political parties and the nationalist parties the Scottish Nationalist Party and Plaid Cymru. Heightened transparency between the nations and regions has prompted debate about the fairness of the geographical distribution of public expenditure and perceived unequal treatment in policy across formerly soft, porous borders within the UK's multinational state (Pike, 2002). Whether such differences become more marked cleavages that test the robustness of the current devolution arrangements might become more apparent as the political ideologies of governments in Westminster and Edinburgh diverge. Unpopular policies may yet stoke the fires of separatism and raise the spectre of the break-up of Britain (Nairn, 2003).

Third, the politics of UK growth will stimulate further research. In contrast to the politics of redistribution inscribed in 'spatial Keynesianism', the current situation is markedly different. Guided by new endogenous economic growth theory, agglomeration is being traded-off against spatial unevenness because of its growth benefits for the national economy (Martin, 2008). The right-wing think-tank Policy Exchange even advocates accelerating the shift to the South-east for the good of the national economy (Leunig and Swaffield, 2008). Anxious not to damage or undermine the growth engine of the UK economy in London and the Greater South-east 'super-region', the UK state appears locked-in to governing uneven development through attempting to address the imbalances and inflationary pressures of geographically concentrated growth with unprecedented levels of public investment.

Last, the recession generated by the financial crisis in 2007–9 has yet to trigger any fundamental reassessment of the role of the state in governing uneven development in the UK. Parts of government have articulated a new 'industrial activism' for the state's strategic role in 're-balancing' the economy in response to the damage and instabilities generated by an over-reliance upon a financialised economy dominated by the City of London (and see Chapter 4). But the more fundamental questions about 'what kind of local and regional development and for whom?' (Pike et al., 2006) have as yet gone unanswered. The prospects for more sustainable and inclusive forms of 'development' – socially and spatially – in the nations and regions of the UK lie uneasily amidst discussion of a 'Green New Deal', reinvigorated neo-Keynesian recovery and public debt and retrenchment.

Further reading

- See House of Commons (2003) on the UK Government's approach to tackling regional disparities.
- For an excellent discussion on devolved governance in the UK see Morgan (2007).
- See also Pike and Tomaney (2008) for further analysis of the governance of economic development in England.

References

Allen, J. and Cochrane, A. (2007) Beyond the territorial fix: regional assemblages, politics and power, *Regional Studies*, 41, 9: 1161–75.

Amin, A., Massey, D. and Thrift, N. (2003) *Decentering the Nation: A Radical Approach to Regional Inequality*. London: Catalyst.

Cooke, P. and Clifton, N. (2005) Visionary, precautionary and constrained varieties of devolution in the economic governance of the devolved UK territories, *Regional Studies*, 6, 39: 437–51.

Fothergill, S. (2005) A new regional policy for Britain, *Regional Studies,* 39, 5: 659–67.

Goodwin, M., Jones, M. and Jones, R. (2005) Devolution, constitutional change and economic development: understanding the shifting economic and political geographies of the British state, *Regional Studies:* 39: 421–36.

Harvey, D. (2001) *Spaces of Capital: Towards a Critical Geography*. Routledge: London.

HM Treasury (2006) *UK National Accounts* (The Blue Book). London: HMSO.

Hogwood, B. (1996) *Regional Boundaries, Co-ordination and Government*. Bristol and York: Policy Press and JRF.

House of Commons (2003) *Reducing Regional Disparities in Prosperity*. ODPM Committee: Housing, Planning, Local Government and the Regions Committee, Ninth Report of Session 2002–03, Vol. 1 Report, HC 492-I. Norwich: The Stationery Office.

Hudson, R. (1989) *Wrecking a Region: State Policies, Party Politics and Regional Change in North East England*. London: Pion.

Jessop, R. (2002) *The Future of the Capitalist State*. Cambridge: Polity.

Leunig, T. and Swaffield, J. (2008) *Cities Unlimited: Making Urban Regeneration Work*. London: Policy Exchange.

Lobao, L., Martin, R. and Rodríguez-Pose, A. (2009) Editorial: Rescaling the state: new modes of institutional–territorial organization, *Cambridge Journal of Regions, Economy and Society*, 2, 1: 3–12.

Lovering, J. (2001) The coming regional crisis (and how to avoid it), *Regional Studies*, 35: 349–54.

Martin, R. (1988) The political economy of Britain's North–South divide, *Transactions of the Institute of British Geographers*, 13: 389–418.

Martin, R. (2008) National growth versus spatial equality? A cautionary note on the new 'trade-off' thinking in regional policy discourse, *Regional Science Policy and Practice*, 1, 1: 3–13.

Martin, R. and Sunley, P. (1997) The post-Keynesian state and the space economy, in R. Lee and J. Wills (eds), *Geographies of Economies*. London: Arnold. pp. 278–89.

Morgan, K. (2006) Devolution and development: territorial justice and the North–South divide, *Publius: Journal of Federal Studies*, 36, 1: 189–206.

Morgan, K. (2007) The polycentric state: new spaces of empowerment and engagement? *Regional Studies*, 41, 9: 1237–51.

Nairn, T. (2003) *The Break-up of Britain: Crisis and Neo-Nationalism*, 3rd edn. London: Verso.

O'Neill, P.M. (1997) Bringing the qualitative state into economic geography, in R. Lee and J. Wills (eds), *Geographies of Economies*. London: Arnold. pp. 290–301.

Pike, A. (2002) Post-devolution blues? Economic development in the Anglo-Scottish Borders, *Regional Studies*, 36: 1069–84.

Pike, A., Rodríguez-Pose, A. and Tomaney, J. (2006) *Local and Regional Development*. London: Routledge.

Pike, A. and Tomaney, J. (2009) The state and uneven development: the governance of economic development in the post-devolution UK, *Cambridge Journal of Regions, Economy and Society*, 2, 1: 13–34.

Rodríguez-Pose, A. and Gill, N. (2005) On the 'Economic Dividend' of Devolution, *Regional Studies*, 39, 4: 405–20.

Rodríguez-Pose, A. and Sandall, R. (2008) From identity to the economy: analysing the evolution of the decentralisation discourse, *Environment and Planning C: Government and Policy*, 26, 1: 54–72.

Tomaney, J. (2000) End of the Empire State? New Labour and devolution in the United Kingdom, *International Journal of Urban and Regional Research*, 24, 3: 677–90.

Wren, C. (2005) Regional grants: are they worth it? *Fiscal Studies*, 26, 2: 245–75.

Acknowledgements

Thanks to Neil Coe and Andrew Jones for the invitation to contribute to this book and their helpful comments. This chapter draws upon research undertaken as part of the UK Spatial Economics Research Centre (SERC) funded by the ESRC, Department for Business, Innovation and Skills, Department for Communities and Local Government and Welsh Assembly Government.

8

HOUSING AND THE UK ECONOMY

Chris Hamnett

AIMS

- To show the role housing and housing finance and wealth play in the economy

- To show the extent to which the UK housing market is a cyclical one and the scale of the current housing market crisis

- To highlight the links between housing wealth, debt and government monetary policy

Traditionally, textbooks on economic geography would rarely have contained a chapter on housing. Historically, economic geography was concerned with the geographical structure of production and production costs, markets and distribution, manufacturing plant size, finance and business services, the service sector, transport costs, the structure of the labour force and costs and so on. While construction was recognised as a component of economic geography, it was and is a relatively small component of output and the labour force, and housing was generally treated as part of consumption and social geography rather than economic geography narrowly defined.

But in fact housing has played a crucial and growing role in the British economy over the last 40 years and the financial and economic importance of housing and the housing market in national economic and financial management has been starkly illustrated from summer 2007 onwards with the emergence of the US sub-prime mortgage lending crisis, the subsequent slump in the housing market and the necessity for government in both Britain and the USA to take over or bail out a number of banks and financial institutions.

It is not an over-exaggeration to claim that the housing market has played a major role in both the economic boom of the last 10 years, and in the subsequent financial collapse and recession. One reason is that the home ownership market sucked in very large amounts of capital, allowed high levels of equity extraction and consumption spending and generated high levels of personal debt. The collapse of the housing market has also intensified the economic recession partly because housing related spending on renovations, fixtures and furnishings accounts for a large proportion of consumer spending and partly because of its impact on personal wealth. When the housing market goes into sharp decline, this has a major effect on consumer spending and the housing market plays an important role in the British economy and financial system. Indeed, Peter Bolton King, chief executive of the National Association of Estate Agents, said in April 2009 that 'the housing market is the engine of the UK economy'. While this is a major exaggeration, there is no doubt that it is an important element.

The chapter is structured as follows. First, it looks at the sub-prime mortgage crisis and the wider financial crisis. Second, it looks at the changing tenure structure of the UK housing market and cyclical nature of the UK ownership market. It then examines four broad areas: first, housing construction and the building industry; second, finance for house purchase and the role of housing wealth and mortgage debt on the economy; third, the role of housing and housing wealth on consumer spending; and fourth, the importance of housing in national economic management.

The sub-prime mortgage crisis and its implications

The problems over American sub-prime mortgages triggered (but did not cause) both the credit crunch and the subsequent financial crisis in summer 2007. What happened was that large numbers of mortgages had been given to low-income households in the USA, often on initial low rates of interest, by mortgage brokers whose main interest was in sales and profits, not repayments. These were packaged and distributed by the banks and mortgage lenders and sold to institutional buyers such as pension funds looking for a steady income stream. When it became clear that many of these households could not, in fact, afford to meet their mortgage payments and mortgage defaults began to rise, the value of many of these mortgage derivatives crumbled, and financial institutions realised that some of their assets were less secure then they had imagined, and the supply of bank credit rapidly dried up. This had disastrous consequences for many major US mortgage finance institutions which had to be either bailed-out or nationalised by the US government in addition to the collapse or take-over of most of the major US investment banks such as Lehman Brothers, Bear Sterns and Merrill Lynch (Bank of England, 2008).

In Britain, this hit many of the former de-mutualised building societies such as the Northern Rock, Bradford and Bingley, and Alliance and Leicester, which had converted into banks and had expanded very aggressively into more speculative mortgages funded not by traditional retail deposits but by loans on the wholesale money markets. When these loans dried up in autumn 2007 as part of the credit crisis it proved impossible for them to refinance their loans and they ran into major financial difficulties. There was a run on the Northern Rock in autumn 2007 as lenders queued up to withdraw their deposits. This led to it being taken over by the government in February 2008. Subsequently, the Alliance and Leicester and Bradford and Bingley were taken over by the Spanish Banco Santander, and HBOS (Halifax Bank of Scotland) was taken over by LloydsTSB as the scale of their financial problems became apparent. In late 2008 the government had to step in to bail out and take share holdings in a number of banks, including the Royal Bank of Scotland and Lloyds/HBOS, as the scale of their problems deepened. While the problem of reckless mortgage lending and the packaging and distribution of mortgage loans worldwide were just part of the wider financial crisis which was based on the over-supply of financial credit and dubious loans, housing finance has played a key role in triggering the current financial and economic crisis (Gowan, 2008)

By 2008 Britain was in a growing housing crisis, driven in part by the unsustainable rise in house prices over the previous 12 years and by the tightening of mortgage loan criteria and the rise in mortgage rates as mortgage lenders dramatically cut back on their lending. From the market peak of c. £200,000 in summer 2007, average house prices had fallen by 20 per cent by early 2009 to £160,000, the volume of sales had halved, as had the value of mortgage loans and the volume of housing construction. Mortgage repossessions rose sharply to 42,000 in 2008, and 48,000 in 2009 and the British newbuild private housing market had slowed to a virtual halt. To understand the events which led up to this crisis, it is necessary to look back at the history of the home ownership market in Britain since the early 1970s and the growing role of housing in the British economy.

The changing tenure structure of the UK housing market

Until the First World War, the British housing market was dominated by private renting which accounted for 90 per cent of households. The interwar period saw the growth of home ownership and also of council renting. By 1945 home ownership accounted for about 25 per cent of households, helped by low interest rates and the growth of the building society movement, but private renting was still by far the largest tenure. Post war, private renting declined steadily until the late 1990's but by 1971 home ownership reached 50 per cent of households and it has now stabilised at just under 70 per cent, with council and other social rented housing accounting for about 20 per cent and private renting for 10 per cent. The growth of home ownership between the two world wars was most marked in South-east England, but it spread across the whole of the UK. Council renting played an important role in the post-war decades, but with the election of Margaret Thatcher and the Conservative government in 1979 the size of the council sector was reduced,

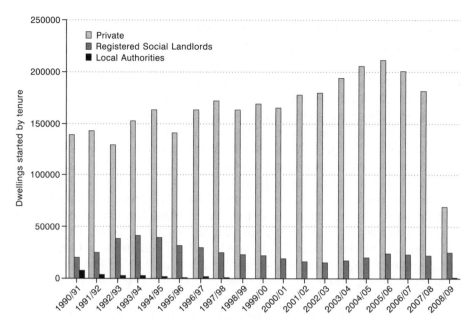

Figure 8.1 House building in the UK, 1990–2009
Source: Table 2.8 House building: permanent dwellings started, by tenure and country, (Department for Communities and Local Government, Housing Construction Statistics, 2009)

particularly as a result of the 'right-to-buy' policy which gave council tenants the right to buy their homes at a large discount to market value. In recent decades there has been virtually no new council house building and social housing has been built by housing associations (see Figure 8.1). The result has been a growing shortage of affordable housing.

Housing in Britain is now predominantly owner-occupied although in the last 10 years, as a result of the rise of assured tenancies and shortholds, there has been a sharp increase in the scale of private renting, particularly of buy-to-let (BTL), the number of mortgages for which had risen from 27,000 in 1998 to over 1 million by 2007 (9 per cent of all mortgages). This proved to be another problem in that mortgage lenders were keen to lend to what was a rapidly expanding market which was focused on new-build city centre apartment developments often bought by speculative investor landlords, some of whom were lured in by prospects of big profits which failed to materialise. When the housing market slump hit, it was particularly marked in these new-build BTL developments, many of which were over valued and had inflated mortgages. Prices have fallen in some cities by over 50 per cent leading to distress sales, large losses and repossessions (Leyshon and French, 2009). Mortgage lenders have become far more cautious on lending to BTL landlords, where evidence of mortgage fraud has come to light and the market is currently in severe difficulties (and see Chapter 5).

The cyclical 'boom and bust' structure of the UK owner-occupied housing market

The UK home ownership market has been shown to be highly cyclical over the last 40 years in that it has had several major house price booms and two major busts. The first big house price boom was 1970–73 when average property prices doubled from about £5,000 to £10,000. There was then a major downturn from 1974 to 1977 as a result of interest rates almost doubling to 15 per cent in an attempt to damp down the economic boom.

Prices then rose sharply again from 1978 to 1980, only to fall back. There was then a major boom in the second half of the 1980s when prices more than doubled until the Chancellor of the Exchequer, Nigel Lawson, raised interest rates to 15 per cent in autumn 1989 in an effort to halt the boom. This led to a major housing market slump which lasted until the mid 1990s and led to over half a million owners being repossessed and perhaps two million households in negative equity (Hamnett, 1993; 1999). From the low point of the early 1990s, average house prices rose 230 per cent to August 2007 when the housing market once again fell off a cliff as the credit crunch took hold. The scale of the boom is shown by the fact that the value of mortgage lending rose from £120 billion in 2000 to £360 billion in 2007 and average house prices rose from £60,000 in 1995 to £200,000 in 2007 (Hamnett, 2009a).

The downturn in the home ownership market began in autumn 2007 as mortgage rates began to rise and mortgage availability contracted as a result of the credit crunch and the government take-over of Northern Rock and the problems facing other mortgage lenders. All the indicators show a sharp fall in housing construction, sales and mortgage lending. Housing sales figures have fallen by some 70 per cent and the number of new dwellings started more than halved from 2007 to 2009 causing major problems for the large housebuilders, many of whom were burdened with debt. Average property prices fell by some 17 per cent in 2008 according to Halifax and the Nationwide building society figures, and national average house prices fell back to the level they reached in 2005. Prices have since recovered somewhat since mid 2009.

The home ownership market in Britain is a very cyclical one, marked by periods of rapid house price inflation, which push house prices up to unaffordable levels, followed by periods of price stability or slumps, in which the overheated market adjusts back to more affordable levels once again. This is seen in Figure 8.2 which shows real (inflation adjusted) house prices from 1975 to 2007 from the Nationwide Building Society. The straight line marks the long-term trend in house prices, which is an increase of almost 3 per cent a year in real terms, compared to the actual pattern of house price change which shows the booms of the early and late 1970s, the late 1980s and the years from 2003–7. Unfortunately, each boom is marked by a correction of greater or lesser severity. The correction of the first half of the 1990s was very severe, and the 2007–9 correction has also been very sharp. While this offers good opportunities for first-time buyers to enter the market it can be very painful for existing owners, some of whom face negative equity or repossession by lenders.

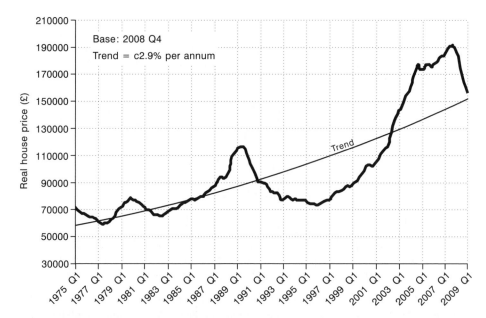

Figure 8.2   Real house prices in the UK, 1975–2007
Source: Produced from data from Nationwide Building Society, http://www.nationwide.
co.uk/hpi/historical.htm, accessed 8 March 2010.

This was a major issue in the early 1990s when 500,000 owners lost their homes through repossessions and 2 million were in 'negative equity' – the value of their house was less than the outstanding mortgage (Gentle et al., 1994). These groups were concentrated among marginal owners, particularly in the South-east, and it is believed that this may have contributed to the Conservatives performance in the 1992 election and losing the election of 1997 (Pattie et al., 1995).

   The cyclical structure of booms and busts in the UK ownership housing market also has a strong geographical component in that each of the booms has started in London and the South-east and then diffused outwards to the rest of Britain. What generally happens is that prices start rising strongly in London and the South-east for 2–3 years (where prices are traditionally much higher), then as price increases slow down in London, prices start to rise in the more peripheral regions of the UK, partly as a result of some owners in the South-east selling and using their increased equity to buy elsewhere, helping to push prices up in turn. In the first 2–3 years of a boom the price differential between London and the rest of the country rises to a peak of about 1.6:1, then falls back.

The building industry, housing finance and deregulation

The most visible link between the housing market and the economy is in terms of house building, and the employment, profits and income that this

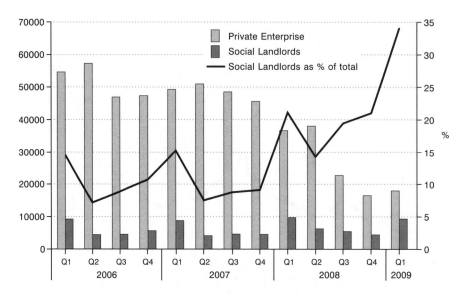

Figure 8.3 Housing Starts, Private and Public, First quarter 2006 – First quarter 2009
Source: Table 2.11 House building: permanent dwellings started and completed, by tenure, United Kingdom (quarterly), (Department for Communities of Local Government, Housing Construction Statistics, 2009)

generates. But house builders have only been building around 200,000 new units a year in recent years compared to the peak of 300,000 in the 1960s. The impact of the credit crunch, falling house prices and the difficulty of getting new mortgages on the house building industry has been dramatic and has led to a large-scale collapse in new building in 2008–9, to half the level of 2007 (see Figure 8.3). First, large numbers of building workers were laid off and a number of major builders and suppliers such as Barratts, Redrow, Taylor Wimpey and Wolsley all saw a dramatic fall in their share price as sales fell and they faced breaching their bank loan covenants. The banks which lent them money to buy land or to takeover other builders when prices were high are also facing very large losses. Second, a large number of estate agents closed as the volume of sales fell. This parallels the experience of the early 1990s (Beaverstock et al., 1992). Third, the volume of mortgage lending fell sharply. Fourth, it has led to a sharp decline in the value of government tax receipts from stamp duty on purchase. Finally, the rise in mortgage rates in 2008 and in mortgage costs for those on low interest deals led to a rise in the number of borrowers in mortgage arrears and in number of repossessions by lenders. The number of repossessions doubled from 21,000 in 2006 to 48,000 in 2009.

A key factor in understanding these trends is that buying a house or a flat is expensive. It constitutes the largest single expenditure most households make. Historically, the average price of property in Britain has varied from 3.5 to 5 times average income. In practice this means that most households are unable to buy a property without taking out a large mortgage to finance the purchase.

In general, first-time buyers have had to seek a mortgage of up to 90 per cent of the purchase price, usually spread over a period of 25 years. The implications of this are that households have to take on a large amount of debt to buy. Assuming, for the sake of simplicity, that the average price of a property is £165,000, and that a first-time buyer puts down a deposit of £15,000, this leaves a mortgage of £150,000. The annual interest on this, at 5 per cent, is £7,500 and it is necessary to pay off the capital sum over 25 years, it is likely that annual repayments will be at least £8,000, most of which will be interest in the early years. Generally, mortgage lenders do not give mortgages of more than four times annual income, which would require an income of £35,000 and repayments could amount to almost a quarter of gross income. Current mortgage interest rates are historically low, and there have been at least three periods in the last 40 years when mortgage interest rates have hit 15 per cent. House buyers are thus heavily exposed to interest rates, and rising interest rates can hit them very hard and make repayment difficult. If buyers lose their jobs they can find it impossible to make repayments and can lose their home if the lender seeks to repossess the property. The volume of secured mortgage lending has grown dramatically since 1995 both in absolute terms and as a proportion of household income. In 2008 the amount of individual debt amounted to £1,500 billion pounds of which £1.225 billion was mortgage debt. The debt to gross domestic product (GDP) ratio rose from around 65 per cent in 1999 to 105 per cent in 2007, the bulk of which was mortgage debt as people have borrowed more to buy their homes. Britain has become a much more indebted society.

The growth of home ownership has led to the development of a large mortgage finance industry whose function is to lend to home buyers. Until the early 1980s, the majority of loans for house purchase were made by building societies such as the Nationwide, Halifax, Alliance and Leicester and the Bradford and Bingley. These had their origins in the nineteenth century and were often locally based, as their names indicate. They traditionally operated conservative lending policies, sometimes lending only 80 per cent of the purchase price and lending a maximum of three times income and requiring potential buyers to save until they had a deposit of 20 per cent. They also raised most of their funds from deposits from savers. This began to change in the early 1980s when the government deregulated mortgage lending and allowed the banks to enter the market and the building societies to access the wholesale money markets for funding to lend (Boddy, 1989).

The result was the rapid growth of mortgage lending which, arguably, helped to fuel the rapid increase in house prices in the late 1980s. A number of large building societies such as Abbey National, Bradford and Bingley, Alliance and Leicester and the Northern Rock also decided to convert from the traditional mutual societies owned by their members into institutions with stock-market listings. The rational for this was that it would enable them to raise money more easily to expand their lending. The conversions were supported by their members, who were offered free shares in the new institutions. The Halifax building society merged with the Bank of Scotland to form HBOS – the largest mortgage lender in Britain. For the first few years all seemed well, and the newly privatised mortgage lenders took a large and growing share of the mortgage market, helped by the relaxation of mortgage lending criteria, with lenders offering mortgages up to 95 per cent of the purchase price and some even offering 100 per cent or even more. Much of the money for lending was raised on the wholesale international money markets which collapsed in late 2007 leading to a financial crisis for many of the demutualised lenders, many of which have since collapsed and had to be bailed out by government or taken over.

In addition, in order to try to increase their market shares, some lenders moved into new markets, such as buy-to-let and self-certified mortgages (where the borrower simply had to declare their income without this being checked). The result was that the traditional conservative lending policies were abandoned and an increasing number of mortgages were given to people who effectively mortgaged themselves to the hilt in order to buy. This proved a problem when house prices began to fall and unemployment began to rise. In addition, it has subsequently become clear that a number of mutual building societies – including the Cheshire, Derbyshire, Dunfermline, West Bromwich, Swindon and Stroud and Chelsea – had either acquired risky mortgage portfolios from secondary lenders near the peak of the boom or made reckless commercial loans. The first three have had to be taken over and the others are facing difficulties from mortgage fraud and bad debts. The result looks likely to be a strengthening of the regulations on mortgage lenders (Hamnett, 2009b; FSA, 2009).

Housing wealth, equity extraction and consumer expenditure

Most house buyers need to furnish their house, and many want to renovate or redecorate, particularly if the house is an older property. This generates high levels of spending on new kitchens, bathrooms and house furnishings such as carpets, tiles and furniture. This in turn has led to the emergence of a large industry catering to house improvement such as B&Q, Homebase, and a large number of furniture and kitchen suppliers such as MFI. The amount of money spent each year on these outlets runs into many billions of pounds, but spending is greatest when households buy or move house. Thus, a big fall in the level of sales when the market slumps as happened in the early 1990s or from 2008 onwards, leads to a sharp fall in housing consumer expenditure with a substantial economic knock-on effect on those sectors catering to housing-related spending.

One of the major consequences of the growth of home ownership over the last 40 years and the associated massive rise in average house prices has been that housing became a major component of household wealth. The total value of housing in the economy rose to £4 trillion in 2007. The growing value of housing equity encouraged many households to withdraw or borrow against some of their housing equity in order to release cash for consumption. This

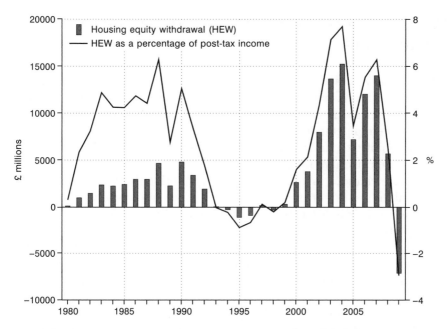

Figure 8.4 Mortgage equity withdrawal (£m and as per cent of post-tax income)
Source: Bank of England, Quarterly statistics on Housing Equity Withdrawal, 2009,
http://www.bankofengland.co.uk/statistics, accessed 8 March 2010.

is possible either by increasing the size of the mortgage on the house and using the money raised for consumer spending, or by taking out a larger mortgage than necessary when moving in order to release some equity. It is also possible to release equity by moving down market to a cheaper property and retaining some of the proceeds. Some mortgage lenders specifically offer equity release mortgages. Architectural writer Martin Pawley (1978) stated that Britain has been playing a game of 'reverse Monopoly', where the aim of the game is not to convert money into houses but to convert houses into money and into consumption.

The scale of equity extraction from the housing market has been very substantial. Bank of England figures on the scale of equity extraction show that it rose to a peak of £12 billion per quarter in 2003: a total of £57 billion for the year. The Bank estimates it accounted for 5–6 per cent of post-tax income and 7 per cent of consumer spending (see Figure 8.4). Significantly, equity extraction went negative in the 1990s housing slump, and went negative again in 2008. Households are not extracting equity from the housing market for other uses but are injecting money to pay off mortgage debt – a total of £5.7 billion in 3rd quarter 2008 and £8 billion in 4th quarter – 3 per cent of post-tax income. The impact on consumer spending is likely to be considerable. Equity extraction from housing has a major impact on consumer spending, both positive and negative, and it arguably reinforces the magnitude of economic booms and slumps.

Housing, the economy and national economic management

When house prices are rising strongly, home owners see an increase in the value of their housing wealth, at least on paper, and this may encourage them to either spend more, to run down their saving or to borrow more and increase their indebtedness in the belief that they have a large cushion of housing equity to underpin their expenditure or borrowing. Conversely, when house prices are falling, and equity is shrinking, home owners may be much more cautious in spending and may increase their savings rate. When prices fall to the extent that many owners move into negative equity, where the value of their house is less than the outstanding mortgage, this may have a sharp effect on consumption. The relationship between house prices and consumer spending is a close one but it is difficult to argue that it is a direct relationship. As Benito et al. (2006: 142) note, although house prices can cause changes in spending, 'the correlation also reflects the influence of common factors like expectations of future income, that affect both house prices and consumption.' The key point, however, is that there is a correlation between house prices and consumer spending. When house prices are rising, consumer spending also rises, and vice versa. Thus, house prices reinforce changes in consumer spending. It is significant that in 2009 the Prime Minister, Gordon Brown, announced a policy to build more social housing in order to support the building industry and generate more spending in the economy.

Housing plays a surprisingly large role in national economic management, because of the very large volume of housing-related mortgage debt and the sensitivity of home owners to changes in the bank rate and mortgage rates. But government often faces contradictory pressures. In periods when the housing market is booming, it may be under pressure to try to take the heat out of the market by raising interest rates and thus deterring buyers from taking on too much mortgage debt. This has happened several times in Britain, first in the early 1970s, when the house price and property boom forced the Chancellor to raise interest rates to 11 per cent in late 1973. A similar event happened in 1978–79. Similarly, the late 1980s boom forced Nigel Lawson, the Chancellor, to raise interest rates to 15 per cent in autumn 1988.

The result was that from the 1989–95 the housing market slipped into a prolonged slump. But at the same time, rising interest rates raise costs for businesses and can damage manufacturing industry, which generally pushes for interest rates and costs to be reduced. As house prices are highest and tend to rise first in the more prosperous London and the South-east, and these increases then get transmitted to the rest of the country, it has been argued by Peck and Tickell (1992) that, if interest rates are pushed up to contain a house price boom, this penalises other sectors of the economy and the housing market in southern Britain may have to be held back at the expense of manufacturing and the North.

In addition, there was a major concern during the 1980s that high and rising house prices in London and the South-east were restricting the free flow of labour from areas of high unemployment to areas of labour shortage. While there were jobs available in the South-east, the level of house prices was such that migration from the northern or peripheral regions was being constrained as people simply could not afford to live in the South-east. Thus, we saw the rise of weekly commuting of builders and others from the North-east to the South-east. Similar issues arose in the 1990s boom regarding the problem of housing key workers in London and the South-east. The level of house prices was such that many workers on low or even moderate salaries could not afford to live in London. Consequently, it proved very difficult to fill a number of vacancies for teachers, and others. There were also problems in high-priced but low-income areas such as Devon and Cornwall, where high income demand from outsiders had pushed up prices to over 10 times local incomes, making it all but impossible for low-income locals to buy. One of the positive effects of the housing market down-turn is that it makes housing somewhat more affordable to lower-income groups (Wilcox, 2006).

Conclusions

Housing plays an important role in economic activity and management, particularly via the finance needed to purchase housing, the booms and busts involved in the market, the wealth it embodies, and the consumption spending it generates. Housing booms help to boost the economy, sometimes to unsustainable levels, and housing slumps contribute to economic downturns. It is important for government to try to keep the housing market stable in order to avoid destabilising booms and busts. The evidence of the last 40 years, however, is that this is very difficult, and the events of the last two years show that the impact of the housing market, and housing finance, on the wider economy can be very severe. By late 2009 the home ownership market appears to have stabilised (at a low level) and house prices seem to have bottomed out and are now rising again, but new construction and mortgage lending is still very depressed and the global recession triggered by the sub-prime mortgage crisis in the USA may have longer to go.

Further reading

- Hamnett (2009a) outlines the geography of house price inflation in London and its links to gentrification and housing affordability.
- Hamnett (2009b) argues that mortgage lenders played a major role in the housing crisis by reckless lending and finance policies.
- Wilcox (2006) documents the problems of housing affordability in contemporary Britain.

References

Bank of England (2008) *Financial Stability Report*, October, no. 24.

Beaverstock, J., Leyshon, A., Rutherford, T., Thrift, N. and Williams, P. (1992) Moving houses: the geographical reorganisation of the estate agency industry in England and Wales in the 1980s, *Transactions of the Institute of British Geographers*, 17: 166-82.

Benito, A., Thompson, J., Waldron, M. and Wood, R. (2006) House prices and consumer spending, *Bank of England Quarterly Bulletin*, Summer, 142-54.

Boddy, M. (1989) Financial deregulation and UK housing finance: government-building society relations and the Building Societies Act 1986, *Housing Studies*, 4: 92-108.

Financial Services Authority (2009) *Mortgage Market Review,* Discussion Paper October.

Gentle, C., Dorling, D. and Comford, J. (1994) Negative equity and British housing in the 1990s: cause and effect, *Urban Studies*, 31: 181-99.

Gowan, P. (2008) Crisis in the heartland, *New Left Review*, 55: 5-29.

Hamnett, C. (1993) The spatial impact of the British home ownership market slump, 1989-91, *Area*, 25: 217-27.

Hamnett, C. (1999) *Winners and Losers: Home Ownership in Britain*. London: Taylor and Francis.

Hamnett, C. (2009a) Spatially displaced demand and the changing geography of house prices in London, 1995-2006, *Housing Studies*, 24: 301-20.

Hamnett, C. (2009b) *The Madness of Mortgage Lenders: Housing Finance and the Financial Crisis*, Institute for Public Policy Research, July. Accessed 8 March 2010 at www.ippr.org.uk/publicationsandreports/publication.asp?id = 664.

Leyshon, A. and French, S. (2009) 'We all live in a Robbie Fowler house': the geographies of the buy to let market in the UK, *British Journal of Politics and International Relations*, 11: 438-60.

National Association of Estate Agents (2009) Response to Budget, 22 April.

Pattie, C., Dorling, D. and Johnston, R. (1995) A debt-owing democracy: the political impact of housing market recession at the British general election of 1992, *Urban Studies*, 32: 1293-1315.

Pawley, M. (1978) *Home Ownership*. London: The Architectural Press.

Peck, J. and Tickell, A. (1992) Local modes of social regulation? Regulation theory, Thatcherism and uneven development, *Geoforum*, 23: 347-63.

Wilcox, S. (2006) *The Geography of Affordable and Unaffordable Housing and the Ability of Younger Working Households to Become Home Owners*. York: Joseph Rowntree Foundation.

9

THE GEOGRAPHICAL PENSION GAP: UNDERSTANDING PATTERNS OF INEQUALITY IN UK OCCUPATIONAL PENSIONS

Kendra Strauss and Gordon L. Clark

AIMS

- To show how three interrelated processes – the balance of public/private sector employment, inequalities in health and longevity, and different financial ecologies – shape the contemporary pensions landscape

- To highlight the necessity of looking beyond the North–South divide in understanding geographical patterns of pension inequality in the UK

- To illustrate the ways in which the processes that shape pension inequality operate across a range of economic scales

Pensions are not a topic that interests most of us until retirement is on the horizon. People of working age are aware that they *should* plan for retirement: a recent survey of attitudes to pensions found that a majority of respondents believed in prioritising saving for the future over current spending (Clery et al., 2007). The same survey, however, also found that respondents were not confident of having enough to retire on, and found pensions confusing to a degree that makes financial planning difficult. In other words, there is gap between British people's perceptions of the importance of pensions, their confidence in the ability of the pension system to deliver a comfortable retirement, and their perceived financial competencies.

Recent research suggests that the survey respondents are right to feel uncertain about the future. Current pensioners are among the cohorts who were of working age during the 'golden age' (the post-war period up to the 1970s) when occupational pension coverage was relatively high and the institutions of the British welfare state were being strengthened and extended. Although slightly lower relative to those of pensioners in other comparable countries, median incomes of the over-65s increased and relative poverty fell between the 1970s and the 1990s, when the percentage of gross domestic product (GDP) transferred to pensioners increased significantly (Pensions Commission, 2004: 130). These intergenerational transfers have tailed off in the last two decades, however, and current workers inhabit an increasingly hostile pensions landscape. Employer (occupational) pension coverage, especially traditional defined benefit (DB) plans which pay a pension based on final salary or a combination of best years and length of service, are in dramatic decline in the private sector and threatened by large unfunded liabilities in the public sector. The state pension has been losing value and is bolstered by means-tested benefits, such as the Pension Credit (which is aimed at the poorest pensioners), and private saving rates are low.

Moreover, it has become apparent that the rise in median income among the over-65s was unevenly distributed, with pensioners in the top quartile achieving the largest increases, and those in the bottom little or no increase (Pensions Commission, 2004: 132). This mirrors widening economic inequality among all sections of the British population, which is producing new geographies of wealth, poverty and financial exclusion in the UK (Fahmy et al., 2008). Previous studies have analysed the geographies of pension consumption and occupational pension inequality (Sunley, 2000), and financial exclusion (Leyshon and Thrift, 1995; Clark and Knox-Hayes, 2007).

This chapter thus seeks to go beyond the analysis of spatial patterns of inequality in pensioner's incomes and assets by mapping three interrelated processes that contribute to the complexity of current pensions landscape. They correspond to the spatial division of public and private sector employment in the UK; the sedimentation of regional health inequalities and differences in life expectancy; and, the evolution of local financial ecologies, which influence individual decision making and attitudes to risk. In so doing it thus seeks to move beyond conception of UK pension inequality through the notion of the North–South divide.

The chapter is in four parts. The first section provides an overview of the British pension system, and defines the geographical pension gap. The second section analyses the processes described above, and the third considers the principal proposed solution to the current pensions 'crisis'. We conclude by suggesting that the proposed national pension savings scheme is likely to fall short of providing a comprehensive solution to the problem of inadequate retirement savings.

The British pension system: an overview

The scepticism and confusion evinced by the respondents in the Department for Work and Pensions' (DWP's) 2006 survey of attitudes to pension saving

(Clery et al., 2007) is unsurprising, considering the complexity of the UK pension system. In Britain pension provision is split between three 'pillars'. The first pillar represents benefits provided directly by the state, such as the British State Pension (BSP). The second pillar encompasses occupational (employer) pensions and covers an array of different pension arrangements, including a state-sponsored earnings related pension (the S2P, or SERPS, which is essentially a state-sponsored occupational scheme), stakeholder pensions, and employer-sponsored plans of both defined benefit and defined contribution varieties. Defined contribution pensions, unlike DB-type plans, pay out a pension based on the amount of contributions and any increase in value due to interest or investment income earned: the amount of the pension is not guaranteed by the employer so the risk lies with the individual employee. The third pillar is made up of private or personal forms of retirement saving, such as private pensions purchased from financial services companies (mostly banks, building societies and insurance companies). Although they have tax advantages, they are not generally provided or subsidised (through, for example, direct contributions) by either the state or an employer.

The majority of current British pensioners derive more of their income from state sources, primarily the state pension, than from private or employer pensions (Pensions Commission, 2004). Successive governments since Thatcher have, however, been attempting to shift the balance away from a reliance on public pensions. The Government's position has been influenced by several factors. First, the British population is ageing and the birth rate is falling. As a result, the ratio of 65+ year olds to 20–64 year olds will increase from 27 per cent today to 48 per cent by 2050 (Pensions Commission, 2004: 4), a trend exacerbated by the ageing of the 'baby boomers'. In 2007, for the first time ever, the population of state pensionable age (over 65 for men and over 60 for women) exceeded the number of children under 16 (Dunnell, 2007). Not only is the population ageing, but people are also living longer. This means that in the future, more people will be drawing their pension for longer. Finally, there is reluctance on the part of the state to fund the rising cost of the state pension system through increased taxation alone, with concerns that intergenerational equality could strain social cohesion, and a general ideological aversion to taxation (even in its most redistributional forms) when compared with other continental European welfare states.

What this adds up to is a policy emphasis on occupational and private forms of retirement saving, an emphasis that belies actual trends in occupational pension coverage. Occupational pension coverage peaked about 1967, when there were 12.2 million active pension scheme members; by 2007 this number had fallen to 8.8 million despite an increase in the number of people in the labour market (ONS, 2008b: 15). The picture is very different in the public and private sectors, however. Public sector membership actually increased during the same period, from 4.1 to 5.2 million, while private sector membership fell from 8.1 to 3.6 million. This is in part of a legacy of de-industrialisation: in the past, manufacturing jobs often had associated pension rights, but the service sector jobs that have replaced them are less likely to come with the opportunity to join a pension scheme (even less a salary-related scheme).

The decline in the total number of active members in the private sector reflects the fall in membership of private sector defined benefit schemes. Defined benefit pensions were traditionally the dominant form of occupational pension in both the public and private sectors but are increasingly being closed or replaced with DC plans in the private sector. The most recent statistics show that these trends accelerated after the millennium, with the number of open private sector schemes (those accepting new members) falling from 54,000 in 2004 to only 28,680, or 56 per cent of the total, in 2007 (ONS, 2008b: 14).

Adding to the woeful picture of private sector occupational coverage is the reality of recent falling returns. The funding position of schemes in the PPF 7800 index – funded defined benefit occupational pension schemes, mostly in the private sector – is highly variable: an overall surplus of £130.4 billion was recorded in June 2007, but by December 2008 this had turned into a deficit of £194.5 billion (ONS, 2009). These trends have also hit DC occupational and private (third pillar) pensions. The latter, however, rather than offsetting the decline in employer pensions, only account for a small percentage of the total and are almost entirely concentrated among higher income groups (due to the tax favoured status of contributions to Approved Personal Pensions (APPs), a form of private pension with flexible investment options).

General pension statistics hint at patterns of inequality engendered by uneven plan membership, the unequal distribution of pension assets, and other related regional population characteristics, but tell us little that is concrete. Unfortunately, ascertaining geographical patterns of pension inequality is not an easy task as most of the pensions data collected by the UK government are not disaggregated by region. Sunley (2000) was able to map patterns of pension exclusion using regional measures of pensioner income and uptake of benefits, and to demonstrate that in the late 1990s employees in Greater London were most likely to be members of an occupational pension scheme (57.7 per cent), while those in the East Midlands (49.2 per cent) were least likely to have an occupational pension (closely followed by Yorkshire and Humberside at 49.7 per cent).

If we look at a general picture of the North–South divide in England a decade later, it is indeed evident that in 2006 more employees in southern England were occupational pension plan members, and more contributed to a personal or private pension plan. If we take the North to include the Northeast, North-west, Yorkshire and Humberside and the East Midlands, the percentage of individuals in the 2006 General Household Survey who reported belonging to any type of pension scheme was 24.4 per cent, compared with 25.9 per cent for the South (which includes the East of England, London, the South-east and the South-west) So there is some evidence that a geographical 'pension gap' does exist. If we disaggregate the two major regions into their

Table 9.1   Pension scheme membership by type, Government Economic Region, 2006

Region	Percentage reported belonging to an occupational pension scheme	Percentage reported contributing to a personal/private pension
North-east	25.9	3.5
North-west	24.1	5.4
Yorkshire and Humberside	24.1	5.1
East Midlands	24.7	5.1
West Midlands	22.5	5.1
East of England	22.6	5.7
London	25.0	4.9
South-east	26.7	6.2
South-west	24.9	5.6
Wales	23.1	4.1
Scotland	30.4	4.0
Britain	25.5	5.0

Note: The survey questions were only asked of those in paid work, including those temporarily away from job or on a government scheme, but excluding unpaid family workers.
Source: Calculated from the General Household Survey, 2006

constituent Government Economic Regions (GERs) (see Table 9.1), however, a slightly more complex pattern emerges.

As Table 9.1 illustrates, by 2006 the percentage of employed persons belonging to a pension scheme had fallen country-wide, with Scotland and the South-east reporting the highest rates of pension plan membership, and the West Midlands and East of England the lowest. Rates of membership of a personal/private pension were also highest in the South-east. Clearly, spatial patterns of pension inequality exist; what drives them is explored in the next section.

Wealth, health and the geographical pension gap

Wealth and employment are two of the most significant variables affecting the accrual of pension assets over time. If we start, in Figure 9.1, by looking at earnings expressed as gross (before tax) annual pay by region, we see that in 2008 London had the highest gross median pay, and Wales and the North East the lowest. As the Pensions Commission (2004: 65) found, private sector employees with higher incomes are more likely to participate in an occupational pension plan: of those earning between £25,000 and £39,999 per annum in 2003, 72 per cent had some form of employer-sponsored pension and 43 per cent were members of a defined benefit (DB) plan, whereas of those earning £9,500–£17,499 only 43 per cent had occupational pension coverage (only 22 per cent salary-related).

But income is only part of the picture. Non-pension wealth is also correlated with pension assets: those individuals with significant pension pots are more likely to have higher levels of non-pension wealth (in the form of, for

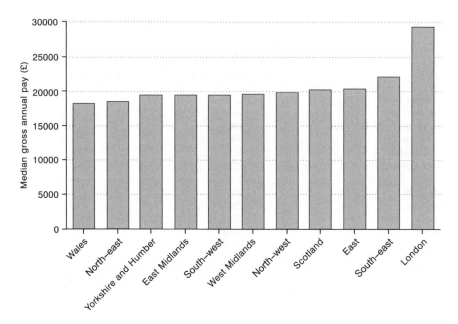

Figure 9.1 Gross median annual pay (£) for all employee jobs by Government Office Region, 2008
Source: ONS (2008) Annual Survey of Hours and Earnings. Newport, South Wales: Office for National Statistics.

example, cash savings, housing assets, stocks and shares, and other financial products). Those with little or no non-pension wealth are also unlikely to have much in the way of savings or investments. So it is no surprise that 2006/07 Wealth and Assets Survey (WAS) data showed that those in the North and Wales were the most likely not to have any accounts or investments, excluding current accounts, or to own property other than their main residence. For example in the North-east, 37 per cent of those over the age of 16 had no accounts or investments, including current accounts, and 9 per cent had no accounts or investments *at all*. In the South-east, on the other hand, only 5 per cent had no accounts or investments – almost half the number in the North-west.

The picture is mixed in London, where there is huge income inequality: at the sub-regional level, Inner London had the highest disposable household income after tax in 2006 (ONS, 2008c), but Greater London also contains some of the most deprived boroughs in England. Thus a full 10 per cent of Londoners had no accounts or investments at all. It is interesting to note that Scotland has relatively high levels of occupational pension membership, but low numbers of people with any accounts or investments. This is where patterns of employment, particularly spatial concentrations of public and private sector work, need to be taken into account.

It is thus necessary to examine the proportion of employees in a region employed in the public sector compared with the private sector. Occupational

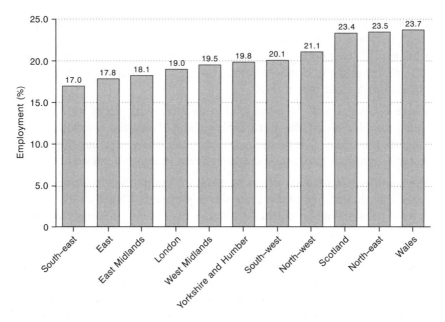

Figure 9.2 Public sector employment as a percentage of all in employment by region of workplace; headcount; average for 4th quarter 2006
Public sector employment estimates for Scotland are published by Scottish Executive (SE) on a quarterly basis back to 1st quarter 1999 from administrative records and surveys of public sector organisations in Scotland.
Sources: Labour Force Survey, returns from public sector organizations; ONS (2006), Scottish Government & Department of Enterprise.

pension coverage is higher in the public sector and, as described above, the majority of public sector plans are defined benefit. As Figure 9.2 illustrates, in the 4th quarter of 2006 public sector employment as a percentage of all employment was highest in Scotland, Wales and the North-east, and lowest in the South-east and East.

This suggests a complex relationship between sector of employment, region of employment, pension plan membership and wealth/ownership of assets. Although there are no available statistics that disaggregate pension scheme membership by sector *and* region, we can surmise from the above evidence that:

- Individuals with higher incomes and more wealth in the form of non-pension assets are also more likely to be members of occupational pension plans and/or hold private/personal pensions.

- Two of the regions with the highest median earnings (the East and the South-east) also have the highest levels of non-pension asset ownership, with lower levels in London reflecting higher levels of income, but also very high levels of income inequality. The South-east also has the second highest level of occupational pension plan membership after Scotland, and the biggest concentration of individuals with private/personal pensions.

- Wales and the North-east have the highest levels of public sector employment, followed by the Scotland. But whereas Scotland has the third highest gross median earnings in the UK, and relatively high levels of asset ownership, Wales and the North-east are at or near the bottom for both median earnings and asset ownership.

- Scotland has greatest percentage of employees with occupational pension plans, the North-east the third highest, and Wales the third lowest. This suggests that there is a complex, rather than linear, relationship between sector of employment, income and pension scheme membership.

- The South-east is second only to Scotland in occupational pension coverage, suggesting that highly paid private sector jobs were almost as likely in 2006 to have pension rights attached to them as well-paid public sector jobs.

While evidence on earnings, assets and sector of employment offer some insights into spatial patterns of inequality in pensions, it is only part of the picture. Regional inequalities in health and mortality affect patterns of labour market attachment and early retirement, and also have equity implications in terms of how pension assets are accumulated and utilised over the life course. As Table 9.2 shows, there are significant regional variations in healthy life expectancy at birth and at age 65. The UK Office for National Statistics (ONS) defines healthy life expectancy (HLE) as the expected number of years an individual can expect to live in good or fairly good self-perceived general health.

Table 9.2   Life expectancy and healthy life expectancy at age 65 by Government Office Region, country and sex, 2001

Government Office Region	Males		Females	
	Life expectancy at birth (years)	Healthy life expectancy at birth (years)	Life expectancy at birth (years)	Healthy life expectancy at birth (years)
North-east	74.5	65.6	79.3	69.3
North-west	74.6	66.4	79.4	69.9
Yorkshire and The Humber	75.4	67.6	80.2	70.9
East Midlands	76.1	69.1	80.5	72.2
West Midlands	75.4	68.1	80.3	71.4
East of England	77.0	71.2	81.4	74.2
London	75.8	68.7	80.8	72.0
South-east	77.2	71.7	81.5	74.8
South-west	77.1	70.9	81.7	74.2
Wales	75.3	66.2	80.0	69.5
Scotland[1]	73.5	65.5[2]	78.9	69.4

Notes: [1] Scotland data are derived from Health Expectancies data for 2001–2003. ONS (2006) Labour Force Survey.
[2] Significantly different from England at the 95 per cent level.
Source: ONS (2006b)

What is striking is that healthy life expectancy for both men and women in the South-east was a full five years higher than in Scotland in 2001. In 2005–7 five of the ten local areas with the highest life expectancy at birth in the UK were in the South-east, while seven of the ten with the lowest life expectancy at birth were in Scotland.

These regional health inequalities have several implications in the context of pensions. First, fewer healthy years means a greater likelihood of periods out of paid employment, and a greater likelihood of early retirement due to ill health. This makes it more difficult to accrue a full pension in a salary related scheme, and reduces the total contributions in a DC scheme. Since ill health is to some extent correlated with poverty, especially in low-income areas, these factors are likely to further disadvantage people who already face significant barriers to building up pension assets. Second, those with DC pensions must buy an annuity (a contract, usually sold by an insurance company, designed to provide payments to the holder at specified intervals after retirement). Annuity rates do not generally take account of local and regional variations in longevity, meaning that someone in inner-city Glasgow (the area with the lowest life expectancy at birth in the UK) will have to pay the same for an annuity as someone in Kensington, which has the highest life expectancy at birth. Poorer people are thus likely to pay more for their annuity contract over the duration of their post-retirement lifespan. The same is true of the state pension: those who accrue full pensions but only live two or three years into their retirement effectively subsidise those who draw pensions for 20 or more years after the age of 65 (for men).

When spatial patterns of wealth, poverty and financial exclusion, employment, and health and longevity disparities are examined, it becomes evident that there is a geographical pensions gap in the UK. What is also clear, however, is that the landscape of UK occupational pensions is too complex to be reduced to a simple North–South divide. Contributing to this complexity are regional variations in attitudes to retirement savings. Previous studies of decision making, attitudes to financial risk and views on financial markets have posited something akin to local cultures or 'ecologies' of finance, which produce and are produced by very different resources, experiences and levels of access to mainstream financial institutions (Leyshon et al., 2004; Clark and Knox-Hayes, 2007). The Department for Work and Pensions survey on attitudes to risk and retirement saving in 2006 (Clery et al., 2007) showed, as illustrated in Figure 9.3a, that few people are confident about their level of pension knowledge, and (as shown in Figure 9.3b) that the majority think a private pension linked to the stock market is too risky. These findings accord with other research that has highlighted a lack of knowledge about pensions and high levels of risk aversion among occupational pension plan members (see e.g. Clark and Strauss, 2008). Moreover, there are geographical differences in perceived knowledge and attitudes to risk.

Those who consider their knowledge to be good or reasonable (Figure 9.3a) were mostly likely to live in the South-west (49.4 per cent, with the most respondents reporting good levels of knowledge) and the South-east (42.2 per cent), while those who characterised their level of knowledge as very patchy

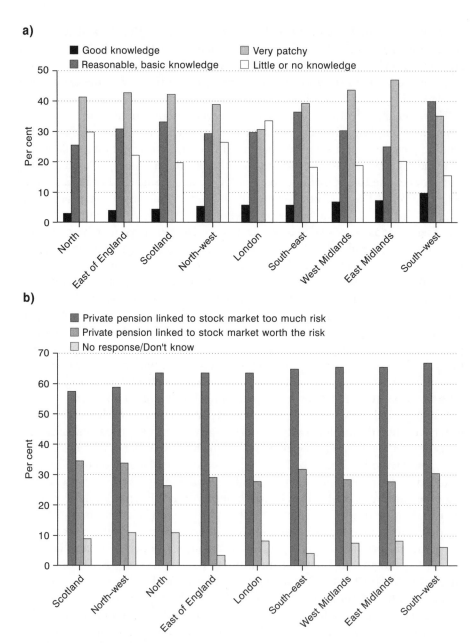

Figure 9.3 (a) Self-assessed pension knowledge and (b) Attitudes to stock market risk, individuals, by Government Office Region, 2006
The North-east and Yorkshire and Humberside regions were combined, and Wales dropped, as these variables had too few cases in the original dataset.
Source: Clery et al., 2007

or little or none were most likely to live in the East Midlands (67.9 per cent) and the North-west (65.5 per cent) with London or the North having the most respondents with little or no knowledge.

When asked about whether they thought 'a private pension linked to the stock market is too much risk – you never know how much the fund will be worth when you retire' (Figure 9.3b), respondents in Midlands and the South were *more* likely to agree than those from the North-west and Scotland, who were more likely to agree that 'a private pension scheme linked to the stock market is worth the risk as it makes the most of your money'. This is an interesting finding, considering the low rates of private pension take-up in the North and Scotland, and seems to indicate that those with lower self-reported levels of financial knowledge are not necessarily less risk averse when faced with decisions about retirement.

Policy responses to the UK pensions crisis

Evidence suggests that the UK pension system is inadequate in a number of ways. Although the Labour government committed to restore the earnings link for the BSP (meaning that the state pension will be linked to earnings rather than prices), the value of state benefits is being steadily eroded while, as illustrated above, occupational coverage is in decline in the private sector and pension assets and entitlements are increasingly unequally distributed. Moreover, all is not well in the public sector. The Government's estimates of public sector pension liabilities (£650 billion in 2006) are contested, with organisations such as the free-market think tank the Institute of Economic Affairs (IEA) claiming that actual liabilities are as high as £1 trillion. Many funded public sector schemes, such as local authority pension plans, were running pension deficits in 2007 that are likely to be significantly worse today. Moreover the political support for public sector pensions, characterised as 'gold plated' in the business and right-wing press, may wane significantly as the 2008–9 recession deepens, especially in areas with relatively low levels of government employment.

In their first report, the Pensions Commission (2004) set out four future scenarios for Britain's ageing population and the pension system upon which it will rely:

1. If the system remains unchanged, pensioners will become poorer relative to the rest of society.

2. To maintain or improve the current levels of pensioner income, taxes/ National Insurance contributions must rise.

3. The aggregate savings rate must rise (e.g. people must themselves save more for retirement independent of state and occupational pension contributions).

4. The average retirement age must rise.

The Commission came out in favour of a combination of measures encompassing options two to four, and the Government has already enacted policies to raise the retirement age for the BSP and discourage employers from setting a mandatory retirement age (which are, paradoxically, especially common in organisations with DB pensions). Other recommendations, aimed at bolstering the BSP, were broadly redistributive: reducing means testing, making entitlement conditional on residency rather than NI contributions and making the BSP universal for over 75s. The Commission's main recommendation for increasing the savings rate, however, which has been taken up by Government, was the establishment of a National Pension Savings Scheme (NPSS). The scheme, as proposed by the Pensions Commission, would automatically enrol all employees not covered by an employer-provided pension and would compel employees to contribute 4 per cent of salary, employers 3 per cent, and the Government 1 per cent. Crucially, the NPSS would be run as a low-cost DC scheme, so the final pension accrued pension pot would depend not only on the amount of contributions, but also on the investment performance of those contributions over time.

Conclusion

Although there has been considerable debate about the shape of the NPSS and the role of private sector financial services companies, the individualised nature of the DC-type design continues to be assumed. Yet as the previous sections illustrate, a DC solution is highly unlikely to compensate for inequalities in income, wealth, health and longevity, and decision-making confidence. Spatial patterns of pension exclusion are thus multi-scalar, encompassing both the macro scales (the national and supranational scales at which policy is formulated and financial institutions and markets operate) and the micro scales (at which individuals make decisions in the context of family and community ties and networks). Only a collective solution embodying the principle of redistribution, which loosens the link between employment and pension accrual and guarantees an adequate income in retirement for all citizens, is likely to produce a radically more equal pensions landscape.

Further reading

- Blackburn (2002) provides a more critical take on the politics of pension policy and the role of finance industry.
- Clark (2003) offers a good summary of the evolution of European occupational pensions.
- Ginn (2003) offers an excellent overview of the issues relating to the gender pension gap.

References

Blackburn, R. (2002) *Banking On Death: Or Investing in Life: The History and Future of Pensions*. London: Verso.
Clark, G.L. (2003) *Pension Fund Capitalism*. Oxford: Oxford University Press.

Clark, G.L. and Knox-Hayes, J. (2007) Mapping UK pension benefits and the intended purchase of annuities in the aftermath of the 1990s stock market bubble, *Transactions of the Institute of British Geographers*, 32: 539–55.

Clark, G.L. and Strauss, K. (2008) Individual pension-related risk propensities: the effects of socio-demographic characteristics and a spousal pension entitlement on risk attitude, *Ageing and Society*, 28: 847–74.

Clery, E., McKay, S., Phillips, M. and Robinson, C. (2007) *Attitudes to Pensions: The 2006 Survey. Department for Work and Pensions Research Report No 434*. Norwich: Department for Work and Pensions.

Dunnell, K. (2007) The changing demographic picture of the UK, National Statistician's Annual Article on the Population, *Population Trends 130 Winter 2007*. London: Office of National Statistics.

Fahmy, E., Dorling, D., Rigby, J.E., Wheeler, B., Ballas, D., Gordon, D. and Lupton, R. (2008) Poverty, wealth and place in Britain, 1968–2005, *Radical Statistics*, 97: 11–30.

Ginn, J. (2003) *Gender, Pensions and the Lifecourse: How Pensions Need to Adapt to Changing Family Forms*. Bristol: Polity Press.

Leyshon, A., Burton, D., Knights, D., Alferoff, C. and Signoretta, P. (2004) Towards an ecology of retail financial services: understanding the persistence of door-to-door credit and insurance providers, *Environment and Planning A*, 36: 625–45.

Leyshon, A. and Thrift, N. (1995) Geographies of financial exclusion: financial abandonment in Britain and the United States, *Transactions of the Institute of British Geographers*, 20: 312–41.

ONS (2006a) Labour Force Survey. Newport, South Wales: Office for National Statistics. http://www.statistics.gov.uk/statbase/Source.asp?vlnk = 358&More = y.

ONS (2006b) Health Expectancies for local authorities in England and Wales, 2001. Newport, South Wales: Office for National Statistics. http://www.statistics.gov.uk/statbase/product.asp?vlnk = 12964.

ONS (2008a) *Life Expectancy*. Accessed 23/03/2009 at www.statistics.gov.uk/CCI/nugget.asp?ID = 168&Pos = 1&ColRank = 1&Rank = 374.

ONS (2008b) *Occupational Pension Schemes Annual Report. No. 15, 2007 edition*. Norwich: Office for National Statistics.

ONS (2008c) *Regional Household Income*. Accessed 23/03/2009 at www.statistics.gov.uk/cci/nugget.asp?id = 1552.

ONS (2009) *Pension Trends – Chapter 12, 27 January*. Norwich: Office for National Statistics.

Pensions Commission (2004) *Pensions: Challenges and Choices*. The First Report of the Pensions Commission. London: The Stationery Office.

Sunley, P. (2000) Pension exclusion in grey capitalism: mapping the pensions gap in Britain, *Transactions of the Institute of British Geographers*, 25: 483–501.

PART 3

LANDSCAPES OF PRODUCTION AND CIRCULATION

10

THE CHANGING GEOGRAPHIES OF MANUFACTURING AND WORK: MADE IN THE UK?

Ray Hudson

AIMS

- To profile the changing role of manufacturing activity within the UK economy

- To evaluate the UK's evolving position within international divisions of labour

- To demonstrate the inability of state strategies to reverse the decline of manufacturing

- To explore the connections between industrial restructuring and uneven development

I wrote this in 2009, in the context of an unprecedented meltdown of the UK economy, in the midst of the deepest recession for 60 years, as a deep and persistent crisis in the financial and banking sector has spread to manufacturing and the 'real economy' of manufacturing and material production (see Figure 10.1). As a result, Marx' evocative phrase that 'all that is solid melts into air' has taken on a renewed and frightening salience at the start of the twenty-first century. This is a truly global crisis but one that is biting deeply and painfully into the fabric of the economy in the UK, providing very visible evidence of the cyclical character of capitalist development. This is a developmental process that is one of crisis prone and uneven development so that growth in some places occurs as others decline, new firms and industries spring up as others shrink or disappear, and as a result there are unavoidable interdependencies between what happens in the economies of different parts of the world. It is also a process characterised by a savage and place-specific

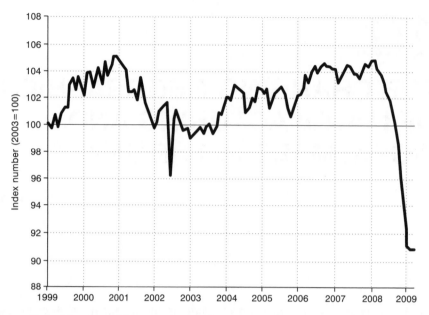

Figure 10.1   Manufacturing output in the UK, 1999–2009

devalorisation of capital, expressed in the closure of factories and workplaces, leading to job losses and unemployment as markets shrink and profitability collapses. Seen against the background of the economic realities of mounting crisis in the UK economy, claims that the policies of New Labour since 1997 have abolished the cyclical character of capitalist development and its swings from boom to bust look remarkably thin and unconvincing (see Chapter 3). The present crisis is particularly savage as it heralds the ending of a booming global long wave rather than just a cyclical downturn in the business cycle. As a result, even (in terms of the logic of capital) highly efficient and productive factories, such as those of Honda in Swindon and Nissan in Sunderland, have had no choice but to cut production and employment in a major way. Given this, it is no surprise that swathes of manufacturing activity in the UK were cut back, with capacity closures and mounting job losses. The crisis in manufacturing has been given an added savage twist as much of the preceding economic boom was fuelled by consumer demand floating on a sea of credit: once it became clear that that credit boom was unsustainable, and banks stopped lending to both companies and consumers, demand shrank dramatically and this was rapidly translated into cutbacks in production, employment and, increasingly, capacity. Or to put it another way, the boom was predicated on claims to surplus-value (that is, the difference between the value of inputs to, and outputs from, a production process – the source of corporate profits) that was yet to be produced, and it became clear that it never was, and never could be, produced in sufficient quantity to ensure that these claims could be met. Once this became clear the edifice on which the boom had been built collapsed as the bubble burst in spectacular fashion – all that had seemed solid did indeed quickly melt into air at a frightening pace with disastrous consequences for the livelihoods and well-being of millions of people.

However, it is important to stress that while the changing structure of the economy reflects the structural contradictions inherent to a capitalist economy, these changes are not simply the result of processes of capital restructuring unfolding automatically and mechanistically. While these contradictions are genetically encoded within the social relations of capital, they become realised in particular ways and places as a result of contingencies and the specificities of places. As such, the ways in which economies, companies and places are restructured reflects the situated practices of a variety of actors – individual, collective and organisational. In particular, changes in geographies of manufacturing production, employment and work in the UK – the what, how and where of change – come about as a result of complex processes of restructuring driven by the competitive strategies of companies that are themselves shaped by the global economic environment. Companies may often be aided and abetted by the actions and policies of national states, with spatially differentiated effects at different scales. These intranational variations to a degree also reflect the differential influence of local (and in the case of Scotland and Wales nationalist) political forces, place-based campaigns to defend jobs and livelihoods and place-based 'growth coalitions' lobbying for investment in new activities, in which trades unions are often involved along with local businesses and local government. However, important though these dynamics of change within the UK are, a key point to bear in mind is that there are interdependencies between what happens within the UK and what happens in the wider world. There are definite relationships between the UK's place in a changing international division of labour and changes in the geographies of manufacturing and work within the UK.

So, before going any further, let me summarise four key arguments and propositions that underpin this chapter. First, capitalism remains – unavoidably – a crisis-prone social system with global reach. Second, the capacity of the national state to cope with such crises is strictly limited. Critically, however, the national state cannot abolish crisis tendencies and uneven development, either temporally or spatially. Third, the economy in the UK has become particularly vulnerable because of the way in which both Conservative and Labour governments over the last 30 years have embraced and promoted a neoliberal regulatory model that reifies the forces of the market, especially those of the global market as an economic steering mechanism. This in turn has altered the position of the UK in the global economy. Finally, a corollary of this has been to reproduce and reinforce uneven development within the UK. One final point by way of introduction: the chapter does not seek not to describe in detail the empirics of change in the UK but rather identifies the key processes that underlie those changes and points to examples of such changes.

Changes in the international division of labour

It is worth reminding ourselves that capitalism is an historically specific form of social organisation that from the outset was constituted as a system that brought different parts of the world into asymmetrical power relationships. In short, it was (and remains) predicated upon the creation of combined and uneven development. Moreover, the UK was central to this process as the original centre of capitalist manufacturing, for a time at least forming the 'workshop of the world'. The International division of labour (IDL) has subsequently undergone a series of transformations, in part enabled by major improvements in transport and communications technologies. Schematically, these transformations can be described as a shift from the old IDL in which manufacturing was concentrated in imperial cores while colonies provided raw material and labour as well as markets for manufacturers, to the New IDL in which parts of manufacturing were selectively re-located to some former colonies (for example, see Fröbel et al., 1980), to the opening up of Central and Eastern Europe following the collapse of state socialism in the USSR (Williams and Balaz, 1999), to the rise of China and India as emergent economic superpowers in an era of globalisation (Engardio, 2006).

For much of the twentieth century the UK was both the origin of outflows of foreign investment and an important destination for inward investment by foreign-based multinational companies, especially those based in the USA. These established branch plants and/or subsidiaries in a range of locations, from around the North Circular road in London and new trading estates such as Trafford Park in the 1920s and 1930s to locations in peripheral regions such as North-east England and South Wales for much of the second half of that century. New factories bearing names such as Hoover, Ford and RCA sprung up in these places and by the 1960s were joined by those of major European multinationals such as Bosch and Phillips. Much of this branch plant investment was linked to central government regional policies, seeking to revive industrial economies in decline (some of the reasons for this decline are discussed below) or encourage the industrialisation for the first time of others, especially in rural areas. More recently, there has been a flow in the reverse direction (see Table 10.1), as multinationals have increasingly re-located routine production from the UK across a whole range of products and parts (in sectors as diverse as clothing, consumer electronics, and autos) – a process with urban and regionally differentiated effects as many of these restructured activities were located in the peripheral regions.

As noted above, these changes were intimately linked to corporate strategies and cannot be properly understood outside of that context as companies restructure as they seek to preserve or enhance profitability – or increasingly in the current crisis, simply survive. Again at the risk of some over-simplification, companies can be thought of as pursuing strategies of 'weak' or 'strong' competition (see Hudson, 2001). In part these changes were linked to strategies of 'weak' competition as multinational companies (both UK and foreign)

Table 10.1   Flows of foreign direct investment
into and out of the UK, 1991–2007 (£million)

Year	Inflow	Outflow
1991	8,418	9,304
1992	8,816	10,107
1993	9,871	17,358
1994	6,046	21,040
1995	12,654	27,604
1996	15,662	21,823
1997	20,296	37,619
1998	44,877	74,159
1999	54,376	124,508
2000	77,029	167,822
2001	36,555	40,844
2002	16,033	33,561
2003	10,276	38,088
2004	30,566	49,713
2005	96,803	44,458
2006	84,855	46,887
2007	91,161	137,678

Source: ONS, MA5 Foreign Direct Investment 2000
and 2007

re-located labour-intensive component production and final product assembly
and fabrication to cheaper labour cost locations or locations with less stringent
environmental and other regulatory requirements as new possibilities opened
up in increasingly globalised markets for production sites. In part, however,
they also resulted from strategies of 'strong' competition, which took two,
sometimes related, forms. First, new fixed capital investment increased the
technical composition of capital in production in the UK, as expressed in the
increased automation of production processes for existing products to increase
labour productivity and output. However, this often led to companies shed-
ding jobs in the process as productivity growth exceeded output growth rates.
Second, it could involve increased research and development activity to create
new products for which companies would have a temporary monopoly in
markets via creating first-mover advantages – this in turn requiring new fixed
capital investment in new production facilities. These changes also resulted
from shifting production out of the UK because of the need to find locations
in which restructured production methods and labour processes, for both old
and new products, could be introduced because of more compliant labour
forces and/or permissive regulatory regimes. In short, to survive as profitable
enterprises in the UK, companies needed to move production up the value
chain – and switch to strategies of strong competition. The alternative was
increasingly to re-locate, especially routine production, outside the UK to
cheaper production-cost locations. This latter tendency has been deeply
reinforced as effective demand has fallen sharply across the global economy,
leading to further reductions in employment and capacity in the UK.

There is a further option open to companies – that is, to shift out of manufacturing and the production of things and into other activities that are seen to offer better possibilities of profit. These have ranged from developing into major aftercare and service providers, so that the production of material commodities such as cars or aeroplane engines simply provides the basis for post-sales services or currency speculation. For example, a company such as Rolls Royce has increasingly come to rely on profits from servicing aeroplane engines rather than fabricating them as a source of profit, while engineering companies in the Tees Valley in North-east England have transformed themselves from metal fabricators into internationally competitive project managers of major construction projects such as power stations and industrial complexes in the newly industrialising emerging economies of the Gulf and Far East. However, such activities are not necessarily insulated from the effects of a crisis that is global in its reach.

A corollary of these changes is that there have been important shifts in what is produced where and by whom, expressed in new patterns of offshoring and outsourcing. Outsourcing refers to re-defining the boundary between 'make, network or buy' and the changing social division of labour – that is, to re-drawing the boundary between those activities that the firm decides to retain within its boundaries and those that it decides to procure from other companies, either as a result of simple market transactions ('buy') or as a result of longer-term collaborative relationships with other firms ('network'). While seeking to reduce cost is an obvious consideration in coming to this decision, companies are also sensitive to retaining key activities and core competences within their boundaries. There are, therefore, strategic limits to outsourcing, irrespective of production cost. To a degree the observed statistical decline in manufacturing employment has been a result of outsourcing, leading to the creation of new sorts of service sector growth, ranging from high order finance jobs in the City of London to back offices and call centres in the peripheral regions, although many of these jobs have disappeared as the ongoing crisis has swept through the banking and finance sectors. However, much of service sector growth is simply an expression of activities re-classified from manufacturing to services and still functionally linked to manufacturing. This suggests a need to think more in terms of linked production systems, constituted via networks of relations across the boundaries of firms – albeit relations with marked asymmetries of power – rather than in terms of manufacturing and service sectors. However, official government statistics on the economy and labour market retain a focus on sectors so that there is a mismatch between concepts and evidence that makes it difficult empirically to track the extent of these reconfigurations of production systems. One implication of this lack of knowledge is that policies may fail to reflect the realities of how companies actually operate and organise production. In contrast to outsourcing, which involves a change in the 'make or buy' boundary, offshoring involves a move of activities outside the UK. This can either be within the same company as part of a changing intra-corporate but international division of labour or as part of an international outsourcing strategy. It is important, however, not to conflate outsourcing and offshoring as there are important differences between them and in their potential implications for activity and employment in the UK.

Finally in this section, it is important to note the crucial significance of national government policy for the economy and its geography within and beyond the UK. These changes in spatial divisions of labour are also a reflection of a continuation of a well-established tendency for national economic policy to focus upon the financial sector and to promoting the UK as a global centre of finance. This reflects the continuing residual emphasis on the City of London as the financial centre of Empire or, to put it another way, it reflects the continuing influence of sections of financial capital and major banks and financial services companies on government policies towards the economy. In practice this has translated into a strongly regionally differentiated policy, favouring the City of London and as a result the Greater South-east. It highlights the point that national polices can – and typically do – have a much more powerful effect on the geography of the economy in the UK than do urban and regional policies that have an explicitly stated objective of seeking to promote greater equality in economic performance and well-being. The crisis that erupted in the latter part of the first decade of the twenty-first century, however, raises serious questions as to the wisdom of this policy emphasis and the prioritisation of the banking and financial services sectors – a point taken up further below.

Nationalisation, rationalisation, privatisation: a new neo-liberal orthodoxy or a prelude to renewed nationalisation?

Until very recently, any talk of nationalisation was regarded as indicative of a nostalgic longing for an era of social democratic policies long-confined to the dustbin of history. With the *de facto* nationalisation of major swathes of the banking sector in 2008 (Northern Rock, RBS, HBOS) and growing calls for government involvement in other sectors of the economy, such as automobile production, this no longer seemed to be the case. The original rationale for nationalisation as part of the post-war social democratic settlement, in strong contrast to the rhetorical emphasis on seizing control of the 'commanding heights' of the economy, was the necessity for the state to step in to socialise risk and underpin profitability in the rest of the economy by securing the future of key industries such as coal, steel and the railways (Hudson, 1986, 1989). One deliberate effect of this was to 'freeze' the economic structures of regions such as North-east England, South Yorkshire, South Wales and West-Central Scotland until the end of the 1950s as new manufacturing industries were prohibited from locating in these regions because they would have attracted men to work in them and away from existing 'heavy' industries. Maintaining employment in coal mining in particular was seen as vital to national economic revival within what was then effectively a single-fuel economy.

By the mid-1970s, however, a new political economy was emerging. This reflected the convergence of a number of inter-related changes. Most importantly, it reflected a growing perception of a need for a changed political strategy from a one-nation to a two-nations strategy – from a social democratic Keynesian welfare state that accorded the state an important role in economic management to a neoliberal workfare state that prioritised markets as economic steering mechanisms. In addition, however, it also reflected changes in global energy markets and pressures to cut back the public sector and reduce the perceived power of its trades unions and open up more space for new foreign capital, again with strongly regionally differentiating effects. In fact this shift was the key to creating the necessary conditions for many of the changes described above. It is also worth noting that while the neoliberal two-nations strategy was developed to its peak as part of the political economy of Thatcherism, in fact its roots can be traced back to policy responses by the preceding 'Old' Labour government to international economic crises in the mid-1970s as it cut back the scope of state expenditure at the behest of the International Monetary Fund as a necessary condition to secure credit (Hudson and Williams, 1995). Nonetheless, Thatcherism took this process to new heights – or depths, depending on how you look at it. It did so in several ways. First, it allowed nationalised industries to source from abroad rather than from each other and so opened their markets to the forces of international competition. Second, it removed capital export controls that had been in place since the late 1940s, thereby opening the door to a shift of UK capital abroad (see Table 10.1) and a loss of hundreds of thousands of jobs from private sector manufacturing in the UK in the early 1980s (Townsend, 1983). Third, it destroyed a similar number of jobs in the nationalised industries as 'rationalisation' led by 'new-style' imported managers, notably Ian McGregor, became a prelude to privatisation – and further losses of capacity and jobs (Hudson and Sadler, 1990). There was a strongly regionally differentiated pattern to these losses. For example, while the South-east emerged more-or-less unscathed from the assault on the formerly nationalised industries, by 2009 almost all of the 400,000 plus jobs that had existed in coal mining, steel making and shipbuilding some 50 years earlier in North-east England had disappeared.

A key point to grasp here is that the Thatcherite two-nations strategy laid the ground for the move from 'Old' to 'New' Labour and the rise to dominance of a new orthodoxy, a dominant neoliberal political economy. Consequently, from 1997 there was no real change in policies towards the economy in general and the former nationalised industries in particular, though there was a gestural politics of concern, perhaps most clearly expressed in the establishment of the Inquiry into the Coalfields. However, this had little practical effect in addressing the problems of lack of work on the former coalfields (Bennett et al., 2000). However, as the financial crisis erupted in 2007–8 the seemingly unthinkable happened again as banks were bailed out on an unprecedented scale and *de facto* nationalised, and as companies across the whole spectrum of the economy became the recipients of state aid in the form of grants and loans. This sudden and dramatic shift in relations between state and economy is indicative of the limits to deregulation and to state disengagement from economy and society if business is to go on 'as usual' and economic growth and capital accumulation is not to grind to a halt.

So what's to be left of UK manufacturing and of manufacturing in the UK?

There is an important distinction to be drawn between manufacturing activity in the UK (which may well be and often is in plants owned by foreign companies) and manufacturing by UK companies (which may well and now typically does take place in other parts of the world). The balance between these two has altered over time as a consequence of the balance of acquisition and merger activity and the flows of manufacturing investment into and out of the UK.

Historically there has been a pattern of UK capital acquiring assets in other countries, with UK companies investing in mines and manufacturing plants of various sorts. While such investment occurred in many parts of the world, initially it was focused on both the formal and informal Empires (the latter referring to those parts of the world that were not part of the political Empire but in which Sterling held sway as the dominant currency). Later, as industrial capitalism developed in North America and Western Europe, these became the main targets for investment by UK firms. More recently, however, there has been a perceptible reversal in foreign investment patterns with substantial swathes of manufacturing in the UK acquired by foreign companies – and not simply those based in the USA and Germany and more latterly Japan, but also companies in emerging centres of power in the global economy such as India (for example, by Tata, with its acquisition of Corus, Land Rover and Jaguar – the latter two from the Ford Motor Company). These changes are starkly indicative of an emerging new order in manufacturing. It is not without irony that this was a change initially encouraged politically by Prime Minister Thatcher in the 1980s as a way of bringing in new managerial and working practices and disciplining labour. This was perhaps most evidently seen in the public courting of Nissan in the 1980s, a move that was above all driven by a desire to expose the managers of manufacturing plants in the UK to new ways of organising production and work and controlling labour at the point of production, as concepts of lean production (aiming to minimise inputs to the production process for a given level of output), just-in-time production (producing in response to demand rather than ahead of it, as in just-in-case Fordist mass production, thereby reducing levels of stock and the volume of work in progress) and new forms of quality assurance and control became the fashion in non-union or company-union dominated factories (Hudson, 1995).

While recognising these inflows, it remains the case that these have been a weak counter-current to much greater outflows of manufacturing investment by both UK and foreign multinationals as the UK economy has again experienced savage waves of deindustrialisation. It has become a deindustrialised – some would say, post-industrial – economy (see Chapter 11). There are those who would claim that there is a newly re-constructed 'knowledge-based manufacturing economy' in the UK – although it is by no means obvious how a manufacturing economy that was *not* based on knowledge could exist in the first place! (For a discussion of these issues, see Hudson, 2009). This new economy is certainly smaller and leaner, focussed on particular places, products and sectors deploying strategies of strong competition – for example, the Oxford to Cambridge (O2C) Arc and the M4 corridor. There is a much greater emphasis upon research and development, design and post-sales service activities, with fabrication a much less significant activity and one that is increasingly outsourced and/or offshored. In these respects, this is certainly a manufacturing economy that depends upon different sorts of knowledges, often produced specifically within higher education institutions in response to demands articulated by companies and combined in new ways.

While often represented as a product of free market capitalism, however, it is important to emphasise that this is an economy that is heavily underpinned by public sector research and development and defence expenditure, heavily concentrated in the South-east, creating markets for both established large companies and new small-and medium-sized companies alike. Once again apparently spatially blind national policies have strongly differentiating effects on the economies of different cities and regions, above all favouring the South-east. Such an economy is largely excluded from the old industrial regions and cities, despite policies to encourage it there, typically linked to the activities of new institutions, notably the Regional Development Agencies (see Chapter 7). For example, in the North-east of England ONE NorthEast has promoted policies to develop new 'knowledge-based' manufacturing in biotechnologies, new and renewable energy and process industries via Centres and Pillars of Excellence, seeking to link firms in the region with the results of research carried out in the region's universities to create new innovative products via the activities of these intermediary institutions (Hudson, 2009). However, despite vigorous efforts, these policies have met with limited effectiveness as they are typically undermined by the effects of global markets and national government policies. While such policies seek to address perceived supply-side shortcomings in these regions, they can do little to address the lack of effective demand in these depressed economies for the products of such new innovative manufacturing firms. As a result, existing patterns of uneven development between and within regions are reinforced rather than redressed.

Table 10.2   Employees in employment in manufacturing, 1979–2009

Year	Number (thousands)
1979	7,053
1982	5,341
1985	4,882
1990	4,605
1995	4,301
2000	4,153
2005	3,290
2006	3,003
2009	2,730

Sources: Hudson and Williams, 1995; Office of National Statistics, Labour Market Trends September 1996 and 2006; Labour Market Report November 2006; Labour Market Statistics Headlines May 2009

Changing geographies of employment, work and worklessness

One consequence of the decline of manufacturing (see Table 10.2: by the end of 2009 manufacturing employment had declined to the level of 1841) and the more general deindustrialisation of the UK economy – and a consequence with profound social effects – is that employment has fallen and unemployment and worklessness have risen in the former manufacturing heartlands. The more recent decline of banking and financial services has exacerbated the problems of the labour market, not least in bringing home the realities of unemployment to many formerly employed in the City of London – by October 2009 the number of people unemployed had risen sharply to almost 2.5 million, a rate of 7.8 per cent, while the number of people claiming Jobseeker's Allowance (the claimant count, that is those eligible for unemployment benefit) by that date had increased to reach 1.64 million, the highest number of claimants since April 1997.

The tensions to which this gives rise have been exacerbated by the growing numbers of workers from other parts of the European Union (EU), especially those countries in Central and Eastern Europe which have recently joined the EU, who have moved to the UK in search of work. In addition there are claims and reports of illegal migrants from other parts of the world (such as Africa and China) employed in the UK (although by definition there is a lack of accurate statistics as to their number). This has led to accusations that these people constitute a new source of cheap migrant labour, at times verging on slavery, pricing UK workers out of jobs in the UK. Put another way, one effect of the presence of these workers is that while commodities may be made and services provided in the UK, they are not necessarily made or provided by UK workers, and this was becoming a source of major tensions with public protests and picketing of workplaces at the start of 2009 in defence of 'British jobs for British workers' – a pointed reference to an earlier statement by Prime Minister Gordon Brown.

There are also important issues of the socio-spatial imbalances between processes and patterns of manufacturing employment loss and new job growth, in both manufacturing and services. At the risk of some over-simplification, there have been significant mismatches between the locations of job loss and new job growth: jobs lost in the formerly industrial North, jobs created in the service-dominated South-east; jobs lost in mining and manufacturing, jobs created in services; permanent and/or full-time jobs lost while casual and/or part-time jobs have been created; jobs lost by older workers, especially men, new jobs taken by younger workers, often women. In part, these mismatches can be seen as a failure of policy in a wide sense (for example, there are issues of education and aspiration among young people that underlie some of these tendencies) but also issues as to what sort of jobs locate where – that is, issues of both supply and demand in the labour market and the way in which this reflects corporate strategies of restructuring, recruitment and 'hire and fire'.

Conclusions – what sort of future economy?

It is clear that the UK economy and economic activity in the UK has undergone tremendous changes over recent decades. While these have been driven by the strategies of private sector companies restructuring in search of profitability – and increasingly survival – it is also clear that the UK Government has been a far from passive actor in these processes. One net result of these changes is that although manufacturing and industry in general remain key sectors of the 'national economy', it is patently obvious that they will no longer provide the route to 'full employment'. More generally, it would seem that the mainstream capitalist economy will lack the capacity to provide sufficient opportunities for employment and gainful work in many parts of the UK which will remain (semi-) detached from the main circuits of growth and capital accumulation. In these circumstances, it would seem prudent to reconsider what we mean by 'the economy' and adopt a more heterogeneous and catholic definition of what counts as part of 'the economy'. More specifically there would seem to be a clear role for the social economy of activities informed by different concepts of value and processes of valuation to those of the mainstream as a way of providing socially useful work and helping maintain social cohesion and political stability – although the social economy itself is unevenly developed (Amin et al., 2002).

Looking to the future, however, there are also the bigger challenges posed by environmental sustainability and the prospect of cataclysmic environmental changes unless processes such as those of global warming are urgently slowed. How can this be made compatible with tackling the fall-out from the global financial crisis (for example, see The Green New Deal Group, 2008) Will a new lean-and-mean manufacturing sector also be green as companies struggle to restore profitability? Will the new 'green' economy be compatible with continuing growth, with technological and other innovations providing technological fixes and solutions to the dilemmas posed by the Laws of Thermodynamics to economic activity, as eco-modernisers optimistically argue? How can a sustainable economy be made compatible with greater social as well as environmental justice? These are some of the bigger questions that face us in the new millennium as we contemplate the sort of manufacturing economy to which we wish to aspire in the UK (and beyond).

Further reading

- Dicken (2007) offers the definitive account of recent changes in the geography of the global economy.
- Hudson (2001) clearly sets out a sophisticated political-economy approach to understanding economies and their geographies.
- Hudson and Williams (1995) provide a theorised account of long-term changes in the UK political economy and its economic geography.

References

Amin, A., Cameron, A. and Hudson, R. (2002) *Placing the Social Economy.* London: Routledge.

Bennett, K., Beynon, H. and Hudson, R. (2000) *Coalfields Regeneration: Dealing with the Consequences of Industrial Decline.* Bristol: Policy Press.

Dicken, P. (2007) *Global Shift: Mapping the Changing Contours of the World Economy,* 5th edn. London: Sage.

Engardio, P. (ed.) (2006) *Chindia: How China and India Are Revolutionizing Global Business.* New York: McGraw-Hill.

Fröbel, F., Heinrichs, J. and Kreye, O. (1980) *The New International Division of Labour.* Cambridge: Cambridge University Press.

Green New Deal Group, the (2008) *A Green New Deal: Joined-up Policies to Solve the Triple Crunch of the Credit Crisis, Climate Change and High Oil Prices.* London: New Economics Foundation.

Hudson, R. (1986) Nationalized industry policies and regional policies: the role of the State in capitalist societies in the deindustrialization and reindustrialization of regions, *Environment and Planning D: Society and Space,* 4: 7–28.

Hudson, R. (1989) *Wrecking a Region: State Policies, Party Politics and Regional Change in North East England.* London: Pion.

Hudson, R. (1995) New production concepts, new production geographies? Reflections on changes in the automobile industry, *Transactions, Institute of British Geographers,* 19: 331–45.

Hudson, R. (2001) *Producing Places.* New York: Guilford Press.

Hudson, R. (2009) From knowledge based economy ... to knowledge based economy? Reflections on changes in the economy and development policies in the North East of England, DOI://10.1080/00343400802662633.

Hudson, R. and Sadler, D. (1990) State policies and the changing geography of the coal industry in the UK in the 1980s and 1990s, *Transactions of the Institute of British Geographers*, NS, 15: 435–54.

Hudson, R. and Williams, A. (1995) *Divided Britain*, 2nd edn. Chichester: Wiley.

ONS (1996/2006) *Labour Market Trends*, September. London.

ONS (2000/2007) *Business Monitor MA5: Foreign Direct Investment*. London.

ONS (2009) *Labour Market Statistics Headlines*, May. London.

Townsend, A. (1983) *The Impact of Recession*. London: Croom Helm.

Williams, A.M. and Balaz, V. (1999) Transformation and division in Central Europe, in R. Hudson and A.M. Williams (eds), *Divided Europe: Society and Territory*. London: Sage.

11

BUSINESS SERVICES: DRIVING THE KNOWLEDGE-BASED ECONOMY IN THE UK?

James R. Faulconbridge

AIMS

- To provide an overview of the nature and role of business services in the UK economy

- To examine the causes and effects of the uneven geographies of business service work in the UK

- To consider the current and future challenges facing UK business services

The role of business services in the UK has been researched and debated by geographers since the early 1980s in the context of wider transformations in the UK's economy. At first, these debates focussed on the rapid growth in employment in services in the UK in the 1980s (see Figure 11.1). In particular, between 1981 and 1987 employment in business services doubled in the UK in the context of only 3.5 per cent growth in all other sectors of the economy. Such growth was initially related to the way manufacturing firms began to externalise, for the first time, key activities such as accounts management, advertising, computer maintenance and repair, legal advice and public relations management. This was intimately connected to the development of what became known as 'post-Fordist' production systems (see Amin, 1994, for more). Manufacturers, instead of employing accountants, advertisers, computer engineers, lawyers and public relations managers in house, relied on a range of external suppliers of such services, thus leading to high levels of employment in specialist service firms.

The emergence in the 1980s in the UK of a significant service sector led to much discussion about both the definition and value of services. O'Farrell and Hitchens (1990) described how, first, debates developed about the

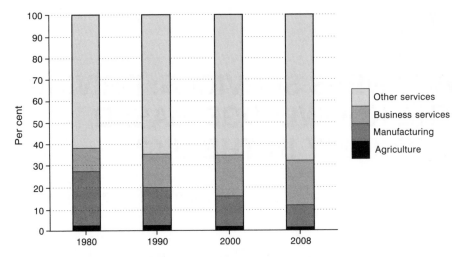

Figure 11.1 Percentage share of employment by sector in the UK, 1980–2008 (Adapted from Office for National Statistics, 2009a)

difference between services provided to 'high street' consumers (e.g. travel agencies and cafés) and those provided to businesses (e.g. advertising and employment law advice). This led to questions about which type of services provided most advantages to the UK economy in terms of growth and employment. Second, the extent to which service industries could exist without manufacturing concerned academics and policy makers. Some suggested that the service economy was exclusively reliant on externalisation from the manufacturing economy and could not exist alone (e.g. Walker, 1985), whilst others argued that the 1980s signalled the start of a UK business service economy completely independent from manufacturing (e.g. Urry, 1987). A further group adopted a mid-range position on the issue (e.g. Wood, 1991). These debates have provided inspiration for much geographical research into the UK business service economy and its role in national and regional economic development over the past 25 years.

In this context, the rest of this chapter examines a number of important questions about the ability of business services to truly drive the UK's economy in the twenty-first century. First, the chapter considers the development of the business service economy in the UK over the past 20 years or so before examining the characteristics of the firms that make up the UK business service economy in the 2000s and the international dimensions to UK business service activity. This reveals the changing forms and geographies of the firms involved in the UK business service economy. Second, the chapter outlines concerns about the geographical unevenness of business service work. The urban geography of the UK business service economy is explained and the causes of geographical unevenness at the national level are outlined with a particular emphasis on explaining the dominance of London in the UK's business service economy. These two issues are then explored in more detail in the last part of the chapter through a case study of legal services in the UK.

The UK business service economy in context

As O'Farrell and Hitchens (1990) highlighted, in the strictest definition, business service firms provide services exclusively to other businesses – they do not provide services to individual consumers in the high street. But there are difficulties in defining business services. Some services (e.g. cleaning, hotels and transport) can be provided both to businesses and individual consumers, causing some slippage in classifications. As a result, the term 'business services', in official data from the Office for National Statistics (ONS) and in geographical debates about services' role in the UK economy, is usually used to refer to only a subset of the services businesses purchase. In particular, it is knowledge-intensive business services (KIBS: sometimes called advanced producer services) that are measured under the category 'business services' in ONS data (classifications L and M in the 2007 Standard Industrial Classification) and which, in policy reports, are said to be vital drivers of the UK's economy. Examples of KIBS are highlighted in Table 11.1 and, as the table shows, these types of services have experienced significant growth (both in terms of number of firms and employment) in the UK over recent years.

Table 11.1   Key business service sectors in the UK economy

Sector	2007 no. of firms	1998 no. of firms	2007 employment	1998 employment
Accountancy	36,000	23,000	245,000	205,000
Law	28,000	23,000	334,000	259,000
Advertising	13,000	11,000	92,000	96,000
Architecture and engineering	66,000	52,000	407,000	328,000
Computing services	109,000	104.000	600,000	442,000
Management consultancy	106,000	46,000	343,000	172,000
Real estate/property management	110,000	60,000	549,000	345,000

Source: Annual Business Inquiry, 2008, www.statistics.gov.uk/abi/default.asp, accessed 9 May 2010

This growing role for KIBS in the UK is a result of the rapidly changing role for manufacturing (see also Chapter 10). As Figure 11.1 demonstrates, throughout the late 1980s and 1990s manufacturing employment in the UK declined rapidly and manufacturing jobs were replaced by business service jobs which showed the greatest growth over the period 1980–2008 (120 per cent compared to only 27 per cent in all other services and a 22 per cent decline in manufacturing). One of the major causes of the decline in manufacturing was competition from developing countries and their low-wage labour forces that provided opportunities for manufacturers to increase profits by globalising and opening factories overseas. In addition, technological advances in manufacturing, such as computer-aided design and machining, reduced the labour intensity of manufacturing and diminished the number of jobs available. As a result, the UK began to emerge as a 'post-industrial society' (Bell, 1973) dominated by service employment and increasingly by business service work not necessarily generated by externalisation from manufacturing. For Allen, writing in 1992, the future growth of the UK economy, therefore, relied on the development of metropolitan service centres in which service firms provided advice to other service firms. For example, advertisers would work for management consultancy firms to promote their services whilst, at the same time, advertisers would buy advice from accountants as part of annual audit processes. As Daniels (1993) described, such a service-based economy began to develop in the UK, centred on London but also other major provincial financial and business service centres such as Birmingham, Edinburgh, Leeds and Manchester.

One defining feature of such service work is that the service provider usually needs to be located in proximity to those buying/consuming the service. Whether it is marketing advice or computer maintenance, the service provider needs to visit, in person, those buying the service, often by attending a face-to-face meeting at the client's offices. Urban agglomeration economies in cities like London and Manchester facilitate this process because when firms locate in close proximity to one another in the same city such meetings can be completed cheaply and conveniently.

In the late 1980s and early 1990s, the role of KIBS in key UK cities, therefore, became the focus of policy initiatives designed to enhance employment in non-manufacturing industry. This was part of attempts to move the UK's economy onto a post-industrial base fit for what became known as the 'knowledge-based economy' in which industries use the knowledge of their workers to produce innovative products. In particular, throughout the 1990s it was emphasised that the UK economy – and in particular the economies of cities like London, Birmingham, Edinburgh, Leeds and Manchester – could benefit from the development of KIBS clusters that serve not just 'local' clients but international markets through exports, thus enhancing growth in gross domestic product (Daniels, 1993).

Business service firms in the UK: key characteristics

The Office for National Statistics (2008a) reports that there are now close to 800,000 firms providing business services. Of these, 98.5 per cent of firms are small enterprises (employing less than 50 employees), with far fewer medium sized enterprises employing 50–249 employees (1.2 per cent) and a tiny proportion of large enterprises employing 250 or more workers (0.3 per cent).

Small and medium sized firms tend to predominantly serve customers on a long-term basis (Daniels and Bryson, 2005). Often the services provided by small and medium sized firms are designed to meet all of a client's needs in relation to one type of business service work. So, for example, small law firms provide advice about everything from employment law and the writing of workers' contracts through to advice about product liability. This helps build long-term relationships with clients and usually means small and medium sized firms have regional reach, serving clients in the same city but also in a larger city-region. For example, Daniels and Bryson (2005) describe how small and medium sized business service firms in Birmingham rely on relationships with clients in the West Midlands area, meaning that at least 25 per cent of their work still involves providing services to local manufacturing firms. But these firms also serve non-manufacturing clients in the larger Midlands region as part of relationships that reach as far South as the outskirts of London, to the edge of the Manchester city-region to the North, and to Nottingham in the East.

Large business service firms share some of the characteristics of their small and medium sized counterparts. Many provide a 'full service' offering and have an important regional client base. But many more large firms than small firms also have a national and international reach, meaning clients are served through multiple national offices within the UK and increasingly through a number of international offices too. This does not mean that small and medium sized firms cannot be national or international in their operations. But it is much less common for a small or medium sized firm to operate outside of their 'local' region.

Research throughout the 1990s demonstrated that the largest UK business service firms – accountants, advertisers, lawyers and management consultants – were increasingly internationalising, in some cases by serving overseas clients from the UK, but more often by opening offices outside the country. As Daniels (1993) highlighted, this internationalisation by the largest firms was designed to allow both existing clients to be served overseas (e.g. when a London-based client of an advertising firm opened an office in Frankfurt their advertisers followed them) and new clients to be identified in overseas markets (e.g. German accountancy firms that might buy advertising services).

The role of large business service firms and their internationalisation was at the centre of important geographical debates throughout the 1990s, both about the relative merits of small versus large business service firms, and the impacts on different regions of the UK of the emergence of international service firms. Some suggested that regional development could be enhanced by the presence of large KIBS firms who had the greatest potential to innovate and develop international markets for services. Others countered this argument with suggestions that small firms could also be innovative and develop crucial regional employment markets (see Bryson and Rusten, 2005, and Daniels, 1993, for summaries of these debates). Whilst the debate is unresolved, the consensus increasingly seems to be that small business service firms are as important as large firms but, as suggested above, large and small firms have different roles with both being needed to make a city's service economy successful.

In this context it is clear, however, that the largest international business service firms which tend to cluster in London are different from many of the other firms operating in cities such as Birmingham, Edinburgh, Leeds or Manchester. In many of the international firms operating in London work is primarily associated with serving clients, such as banks, which are part of the international financial system. As a result, these firms have almost completely disconnected themselves from a system in which business services serve manufacturing. Some of these international business service firms in London employ in excess of 1,500 workers in their offices and have turnovers exceeding £1 billion per annum, mainly because of the demand for advice from clients such as banks both inside and outside of the UK. As a result, a disproportionate amount of the business service exports generated in the UK – estimated at £40 billion in 2008 (Office for National Statistics, 2008b) – originate from London, and it is also London-based firms that dominate business services employment in the UK, with London and the South-east accounting for 44 per cent of total employment in the sector. But, as the figure of 56 per cent of UK business services employment outside of London and the South-east suggests, many types of business service work in the UK are not London-based and, as Table 11.1 demonstrates, do not occur in large firms. This has led to concerns about economic policies that focus on the role of the London-based business service economy at the expense of the business service economy that operates in other cities throughout the UK. It is to these geographical debates about the unevenness of the UK's business service economy that the chapter now turns.

Geographies of the UK business service economy

As already noted, in the UK there is a markedly uneven geography of business service work. In 2003, for example, KIBS accounted for 38 per cent of all employment in London in 2003 but only 16 per cent in Leeds and Manchester, 13 per cent in Birmingham and 12 per cent in Liverpool (Wood, 2006). Other UK cities have even lower levels of business service employment. Indeed, as Wood (2006: 354) argues, despite an evermore important role for KIBS in a number of UK cities, 'whatever benefits this bestows on Birmingham or Manchester, by far the richest social, economic, and cultural environment for KIBS development in the UK remains London.' The City of London has, therefore, been held up as a role model, demonstrating the benefits other cities can reap from a healthy KIBS sector. Indeed, there is no shortage of government publications celebrating the success of the City, most notable of which are those produced by the Corporation of London. As they chime in their latest report on the City's contribution to the UK economy:

The growth of London's financial and related business services cluster has coincided with a big rise in the contribution these sectors make to the UK economy as a whole – financial and professional business services now generate around 13 per cent of GVA [gross value added]. As they have become an increasingly important source of value-added for the UK, so too the success of London's financial and related services industry has become increasingly important to the overall success of the UK economy. (The Corporation of London, 2004: 5)

Such celebratory language hides, however, two significant messages about the role of KIBS in the UK economy. First, other cities are reminded that they cannot blame their lower level of knowledge-intensive business service employment on London's success. As an earlier report on London's linkages to the rest of the UK economy suggested:

London's growth has neither been at the expense of the rest of the UK, nor is it likely to be in the future. While employment in London has risen by 645,000 over the last ten years, employment in the rest of the country has increased by 2.3 million. Employment growth outside London has been nearly 2.5 per cent faster over the last ten years when London has been expanding than it was over the previous ten years when employment in London shrank. (The City of London, 2008: 13)

Second, and related to the first, the implicit message in such reports is that other cities should 'catch up' and strive to increase the role of KIBS in employment. As Massey (2007: 104–5) reveals, 'An insight into this mind-set can be derived from an opening statement of position [from the Department for Trade and Industry]':

The high degree of persistence of regional differentials points to significant problems in under-performing regions. Growth theory would suggest that market forces should result in convergence of regional GDP per capita over time. A lack of convergence in the medium to long term is therefore a likely indicator of series market failures in a number of UK regions or countries. (HM Treasury/DTI, 2001: 3)

Cities such as Birmingham and Manchester should, then, according to government policy, pull themselves up by their bootstraps so as to compete for UK KIBS jobs. But this fails to recognise the effects on consumers of business services of media and government reports that celebrate the success of KIBS in London. The constant celebration of the City's success leads to a belief that, in one way or another, only in the City of London are the UK's *most* advanced and innovative business services to be found. As a result, business services outside of London are portrayed as less innovative and of lower value.

Of course, it is not the case that KIBS only exist in London. As Bryson and Rusten (2005) argue, important and innovative services are at the heart of successful urban economies outside of major world cities like London. But the relative concentration of KIBS in London, reinforced by the impression created by the media and policy documents that other cities lack innovative KIBS, has major implications for cities such as Birmingham, Manchester and Leeds and attempts to grow the role of KIBS in these cities. This suggests, then, that the role of business services in the UK's economy is more complex than it first appears. The nature of firms in the business service economy and the geography of work have important local/regional/global dimensions that need to be examined. In addition, the geographical unevenness of business services mean not all parts of the UK benefit equally from the high wages and employment associated with KIBS. In the next section, these issues are examined further through a case study of legal business services.

Case study: legal business services

The UK's legal system makes a distinction between England and Wales, Scotland and Northern Ireland and requires practitioners and firms to register with Law Societies representing each of the jurisdictions. As a result, lawyers practise within one of the jurisdictions only. In this case study the English legal profession is focussed upon.

There are over 13,000 law firms registered with the Law Society in England and Wales with a large proportion offering services to business in one form or another. But the geography of UK legal services in many ways reflects the UK geography of business services. As a result, cities such Birmingham, Leeds, Manchester and, in particular, London all have disproportionately important roles as centres for the production and delivery of legal business services. As legal services are subsumed within the broader categories of business and financial services in national statistics for the UK, it is difficult to obtain precise data for employment in each city. Membership of a local Law Society does, however, give some indication of the number of lawyers in each city. The City of London Law Society represents 17,000 lawyers practising in the 'square mile' of the City, whilst Birmingham's Law Society has 2,500 members and Manchester's 1,800.

It is immediately clear, then, that London dominates in the UK as far as legal business services are concerned. This is a result of London's role as a world city, which means that both a disproportionate amount of finance-related work takes place in London and international transactions associated with the activities of global firms get channelled through the City. But it is also clear from existing studies that Birmingham, Leeds and Manchester have an important role in the provision of national and international corporate legal services. As Leyshon et al. (1989) noted, the three cities absorb work displaced from London because of the city's high fee levels and the inability of firms in London to meet all of the demands of UK clients. Indeed, as provincial financial centres, firms operating in the three cities offer services

Table 11.2 Selected UK-based law firms taken from *The Lawyer's 2006 UK 100 Survey*

Firm	Place in top 100	Turnover (£m)	Profit margin (per cent)	Number of overseas offices
Clifford Chance	1	914	27	27
Eversheds	7	*302.8*	*19*	*18*
Clyde & Co	21	104	38	11
Irwin Mitchell	22	*102*	*23*	*2*
Shoosmiths	34	*60.8*	*19*	*0*
Reynolds Porter	40	51.8	36	0
Pannone & Partners	60	*33.5*	*30*	*0*
Brabners Chaffe Street	97	*17.1*	*43*	*0*

Note: Firms in italics have offices in at least one of the major cities for business services outside of London. Other firms are solely London-based. Firms were purposely sampled to represent organisations with high, medium and low levels of turnover within *The Lawyer's* survey.

ranging from relatively low-skilled advice (e.g. about employment contracts) through to knowledge-intensive advice (e.g. relating to merging two firms). But, what other effects does the dominant role of London in the geography of legal business service work have on 'provincial' centres? To help understand this issue it is worth examining a range of different UK law firms and their work in London, Birmingham, Leeds and Manchester.

Table 11.2 provides a selection of the firms in *The Lawyer's UK 100 Survey* of leading law firms. It was deliberately constructed to include both the largest and smallest firms and firms operating in London but also provincial financial centres. As is clear from Table 11.2, the top 100 firms in England are a diverse collection with those included having turnovers ranging from £914m to £17m. None of these firms are insignificant enterprises, but the scale of their operations does vary dramatically. In addition, Table 11.2 also reveals some other interesting trends worth considering further in discussions of the knowledge intensity and competitive advantage of firms in the UK.

First, while London, Birmingham, Leeds and Manchester are all dominated by small and medium sized firms, London does stand out as being different in one important way: there are roughly double the number of very large firms (employing 250 or more people) compared with the other three cities (ONS, 2008a). This is reflected in *The Lawyer's* survey, which shows that all of the firms operating in Birmingham, Leeds and Manchester have modest turnovers when compared to their London counterparts. For example, Eversheds, the largest firm in Table 11.2 operating in all four cities, has a turnover of less than one-third of the largest UK firm, Clifford Chance, which only has offices in London. In terms of profitability, a similar trend is apparent. Firms outside of London are generally much less profitable than their London-based equivalents. This is true for several of the firms in Table 11.2 and profit margins for firms with offices outside of London in *The Lawyer's UK 100 Survey* are, on average, 6 per cent below those of firms operating only in London. But how do we explain this?

One reading of this trend relates to the innovativeness, or at least the perceived innovativeness, of legal service firms operating in London. Clients generally expect to pay more as the services they buy are, in their eyes at least, more unique, innovative and designed to deal with complex problems. Profit margin can, therefore, be taken as a crude indicator of knowledge intensity. This suggests that a significant proportion of the work taking place outside of London is less knowledge-intensive than that occurring in London, as Wood (2006) and others have argued. However, there are examples of firms in Birmingham, Leeds and Manchester generating higher profit margins than similar London-based rivals. Both Pannone & Partners and Brabners Chaffe Street achieve higher profitability than London-based rivals of similar sizes, reminding us that stereotypes of all firms being more or less knowledge-intensive in London and 'other' cities respectively is too simplistic. Yet there seems to be a clear trend in which, overall, the very large firms in London out-compete their 'regional counterparts' and have a disproportionate share of knowledge-intensive legal work. In order to further explain this, an additional factor needs to be taken account of.

Table 11.2 also reveals that firms operating outside of London are, in some instances, connected into global business networks and generate legal service exports. *The Lawyer's UK 100 Survey* shows that 35 per cent of the firms operating outside of London have overseas offices. But 65 per cent of firms operating only in London have overseas offices. This suggests, as work on world city networks generally has indicated (Taylor, 2003), that cities such as Birmingham, Leeds and Manchester have important international connections but London has more intense connectivity to other world cities. Perhaps one of the defining features of international work is its complexity and, therefore, the knowledge intensity of the services needed to make it possible. The over representation of international work in London and the disproportionate amount of the £3 billion of the UK's legal exports every year generated in London further suggests, therefore, that London has a competitive advantage in terms of UK knowledge-intensive legal business services. But should we be worried about this uneven geography?

At its simplest, the uneven geography of legal KIBS mirrors the trend in other business services and means that there is a disproportionate concentration of the best paid jobs in London, thus sustaining a North–South divide in terms of average incomes and wealth. Importantly, though, these inequalities are self-reinforcing. Birmingham, Manchester and Leeds suffer from a 'London effect' on their work, with clients making a distinction between, very broadly, super profitable, innovative and international London-based firms and less profitable firms in Birmingham, Leeds and Manchester. Clients are seduced by the rhetoric of both London-based firms and promotional materials such as those circulated by the Corporation of London which lead to an ingrained belief that knowledge-intensive business service work can only be completed in London. Similarly, it is hard to tempt the most talented lawyers to cities like Birmingham, Leeds or Manchester, again because of the power relations associated with London's reputation as the leader of UK business service work. London effectively acts as a vortex, sucking in the best graduates and qualified lawyers who, just like clients, tend to believe that the only place to work if you want to engage in cutting-edge legal work is London. This state of affairs very much reflects Massey's (2007) argument about the power relations that support London's pre-eminence in the UK and which lead to uneven development with cities such as Birmingham, Leeds and Manchester constantly struggling to 'catch up' with London.

Conclusions

So what does the discussion above tell us about the role of business services in the UK's knowledge economy? First, it is clear that KIBS are important in both lubricating the business activities of a range of manufacturing and non-manufacturing UK-based firms and also in the development of a post-industrial service economy that creates employment, economic growth and significant exports from the UK. But KIBS are distributed in a geographically uneven way, meaning London and the South-east benefit disproportionately from the employment and wages generated. So are business services really driving the knowledge economy for all of the UK's residents? Questions seem to exist about the extent to which policies have, to date, effectively created the conditions for the development of a truly *national* knowledge-intensive service economy.

In addition, the financial crisis and recession of 2008–9 raised further questions about the role of business services in the UK's economy. The recession led to a 3.25 per cent decline in the number of business service jobs in the UK (Office for National Statistics, 2009b). This is unsurprising. The dominance of non-manufacturing, finance-related business services in the UK means the business service economy was affected severely by the crisis that hit financial markets in 2007 and 2008. So an economy that is heavily reliant on business services for employment did not prove to be recession-proof and whilst some business service firms benefited from the recession (e.g. lawyers specialising in bankruptcy law), most were hit by massive drops in demand for their services. Also unsurprising was the fact that the big winner in the late 1990s and early 2000s from KIBS and services disconnected from manufacturing, London, was affected more by the financial crisis than cities such as Birmingham, Leeds or Manchester because of its reliance on work generated by financial markets which effectively closed in late 2008 and early 2009. This led the *Financial Times* to proclaim 'It's grim down South' in a jibe at London's disproportionate share of UK job losses in 2008 (around 25 per cent of all losses). The North–South divide in relation to business services closed a little then, as a result of the financial crisis. But it still exists and has major effects on the lives of many people.

All in all, then, it seems that there is a key role of business services in the UK's knowledge economy, but that a fuller understanding is needed of the diversity of business service work, the impacts of geographical inequalities in KIBS and the long-term sustainability of an economy based on business services.

Further reading

- Bryson et al. (2004) offer a comprehensive and in-depth look at the nature, geographies and role in the economy of business services.
- Sassen (2006) is the definitive text analysing the characteristics and role of global cities such as London in international business service work.
- Wood (2006) provides an interesting and data-rich analysis of the geography of knowledge-intensive business service work in the UK.

References

Allen, J. (1992) Services and the UK space economy: regionalization and economic dislocation, *Transactions of the Institute of British Geographers*, 17: 292–305.

Amin, A. (1994) *Post-Fordism: A Reader*. Oxford: Blackwell.

Bell, D. (1973) *The Coming of Post-Industrial Society*. London: Heinemann.

Bryson, J.R. and Rusten, G. (2005) Spatial divisions of expertise: knowledge intensive business service firms and regional development in Norway, *The Service Industries Journal*, 25: 959–77.

Bryson, J., Daniels, P.W. and Warf, B. (2004) *Service Worlds*. London: Routledge.

City of London, The (2008) *Aviation Services and the City*. London: City of London.

Corporation of London, The (2004) *London's Linkages with the Rest of the UK*. London: The Corporation of London.

Cohen, N. (2009) Services sector returns to growth, *The Financial Times*, 4 June.

Daniels, P.W. (1993) *Service Industries in the World Economy*. Oxford: Blackwell.

Daniels, P.W. and Bryson, J.R. (2005) Sustaining business and professional services in a second city region, *The Service Industries Journal*, 25: 505–24.

HM Treasury/DTI (2001) *Productivity in the UK 3 – The Regional Dimension*. London: HM Treasury.

Lawyer, The (2006) *The UK 100*. London: The Lawyer.

Leyshon, A., Thrift, N. and Tommey, C. (1989) The rise of the British provincial financial centre, *Progress in Planning*, 31: 151–229.

Massey, D. (2007) *World City*. Cambridge: Polity.

O'Farrell, P.N. and Hitchens, D.M. (1990) Producer services and regional development: key conceptual issues of taxonomy and quality measurement, *Regional studies*, 24: 163–71.

Office for National Statistics (2008a) *UK Business: Activity, Size and Location – 2008*. London: Office for National Statistics.

Office for National Statistics (2008b) *United Kingdom Balance of Payments: The Pink Book*. London: Office for National Statistics.

Office for National Statistics (2009a) *Labour Market Statistics (First Release)*. London: Office for National Statistics.

Office for National Statistics (2009b) *The Impact of the Recession on the Labour Market*. London: Office for National Statistics.

Sassen, S. (2006) *Cities in a World Economy*, 3rd edn. London: Sage.

Taylor, P. J. (2003) *World City Network: A Global Urban Analysis*. London: Routledge.

Urry, J. (1987) Some social and spatial aspects of services, *Environment and Planning D: Society and Space*, 5: 5–26.

Walker, R.A. (1985) Is there a service economy? The changing capitalist division of labor, *Science and Society*, 49: 42–83.

Wood, P. (1991) Conceptualising the role of services in economic change, *Area*, 23: 66–72.

Wood, P. (2006) Urban development and knowledge-intensive business services: too many unanswered questions? *Growth and Change*, 37: 335–61.

Acknowledgements

Some of the ideas developed in this chapter emerged from work funded by the Socio-Legal Studies Association and completed with Daniel Muzio.

12

AGRICULTURAL RESTRUCTURING AND CHANGING FOOD NETWORKS IN THE UK

Brian Ilbery and Damian Maye

AIMS

- To provide an introductory insight into the major phases of agricultural restructuring in the UK and their geographical ramifications

- To examine how the main processes of change have affected the wheat and potato sectors

- To explore the emergence of alternative food networks in the UK

- To highlight the possibility of a third phase of restructuring in twenty-first century UK agriculture

As with many other economic sectors, UK farming was affected by the economic downturn of 2008–9 and especially by volatile prices for some agricultural commodities (e.g. wheat), feedstuffs and energy. This coincided with governmental and scientific views that the environmental impacts of modern agriculture can no longer be ignored. Indeed, an influential report on the future of food and farming in the UK (Chatham House, 2009: 5) suggested that 'a system that is able to reconcile the often conflicting goals of resilience, sustainability and competitiveness and that is able to meet and manage consumer expectations will become the new imperative' (see also Bridge and Johnson, 2009). This could be interpreted as the start of a 'neo-productivist' phase of agricultural restructuring in the UK, characterised by an effective, rather than exploitative, use of resources over a long time period. In response to the depletion and rising costs of energy supplies, population growth and climate change, neo-productivism requires a lower carbon–lower input system as well as 'technological innovation and the transfer of best practice'

(2009: 5); the debate on genetically modified (GM) technology may thus have to be reopened. However, it is still far from clear what the main dimensions of agricultural neo-productivism will be.

This chapter is structured into four main sections. After a brief conceptual insight into the main phases of agricultural change in the UK, the second section provides an overview of agriculture in the UK today before discussing some of the main features of agricultural restructuring in the UK arable sector. The third and fourth sections focus on 'alternative' food networks and food security, including the GM debate. A summary of the overlapping, co-existing and geographically varied nature of agricultural restructuring and food networks in the UK is provided in conclusion, together with a warning not to simply proclaim a new era of agricultural neo-productivism.

Conceptualising agricultural restructuring in the UK

Initial stages of agricultural restructuring in the UK, from the early-1960s onwards, led to both the modernisation and industrialisation of farming (see Ilbery and Bowler, 1998 for details). Modernisation has been characterised by three key processes:

- *Intensification*, whereby output per hectare has increased, aided by mechanisation, greater use of chemicals (pesticides, fungicides and fertilisers) and disease-resistant varieties.

- *Specialisation*, where less profitable farm enterprises have been eliminated in favour of fewer products.

- *Concentration*, where output and resources have become concentrated on fewer farms and in specific regions.

In turn, modernisation led to the industrialisation of farming, characterised by both the adoption of many practices from the manufacturing industry, such as product and labour specialisation and assembly-line type production systems, and the increasing integration of agriculture into the food supply chain. The food chain (or food network) concept helps to trace the routes taken to get particular foodstuffs from 'farm to fork'. Geographers have used it to literally 'follow' the nature of connections from the supply of farm inputs, through the farm production process, to food processors, distributors, retailers and consumers. Some prefer the idea of a network because it disturbs the linear exchange processes inferred by the food chain concept; indeed, the way that food travels from farm to fork is not always straightforward. However, policy analysts tend to use the food chain concept, even if it does not capture the complexity of contemporary food provisioning (Jackson et al., 2006). In industrialised farming systems, the emphasis switches away from conditions on the farm and towards the relationship between farming and different types of industry. 'Vertical integration' is a term used to describe the control exerted by one large (often multinational) company over most parts of the food supply chain, from the supply of farm inputs (e.g. feed and seed) and the farming of the land to the processing, storing, packaging and marketing of produce, and the distribution and retailing of food.

Modernisation and industrialisation characterised UK agriculture from the 1960s until the mid-1980s, in what has been referred to as a phase of agricultural 'productivism'. Farmers took an exploitative rather than conservative attitude towards their natural resources and aimed to produce as much food as possible at minimum cost. This process was spatial in nature, with productivity increasingly concentrated in favourable production areas such as East Anglia and south-west England for arable and livestock farming respectively. As a consequence, soils were eroded, hedgerows removed, wetlands drained and watercourses contaminated by agrochemicals. Concerns also began to emerge over a series of food scares such as E. coli, bovine spongiform encephalopathy (BSE), foot and mouth disease (FMD) and bird flu. From the mid-1980s, in response to the environmental disbenefits and unstable nature of the industrial model of farming, attempts were made to reduce farm output and to integrate farming within broader economic and environmental objectives, in what became known as a phase of 'post-productivism' (Wilson, 2001). A movement towards more extensive and diversified farming systems was encouraged by government policy through, for example agri-environmental programmes and grants to help farmers diversify into non-farming activities (e.g. tourism and recreation) on their farms. Also, consumer demand for more traceable, higher quality, organic and 'local' food increased and attempts were made to encourage farmers to process and/or market their produce directly to the consumer through outlets such as farmers' markets, farm shops and box schemes.

Research by geographers has shown that the two farming systems (productivist and post-productivist) tend, in reality, to co-exist and Ward et al. (2008) have warned against seeing agricultural restructuring as a simple transition from productivism to post-productivism. Instead, they advocate a commodity-specific approach to agricultural restructuring, as adopted in the next section. This suggests that high input–high output farming is still dominant and spatially differentiated; while the core farming areas have remained essentially productivist, more marginal farming areas have attempted to embrace agri-environmental and diversification schemes, and to focus on providing 'local' and specialist foods for niche markets. Of course, examples of farm diversification and local food systems can also be found in core farming areas, just as examples of intensive farming systems are identifiable in more marginal areas. It is highly likely, therefore, that elements of both productivist and post-productivist farming systems will be contained within emerging conceptualisations of neo-productivism, the contours of which are still to be drawn.

Agricultural restructuring in the UK arable sector

Agriculture accounts for less than 1 per cent of the UK's gross domestic product (GDP), but this rises to 7 per cent when the whole food chain (production,

Table 12.1   Agricultural land use and
livestock in the UK, 2008

	Thousand hectares
Total arable crops	4,569.8
Of which:	
Cereals	3,274.3
Oilseeds	620.6
Horticulture	170.1
Potatoes	143.6
Total permanent grassland	10,394.9
Total temporary grassland	1,137.1
Woodland	606.2
Total agricultural area	17,463.8
	Thousands
Total poultry	166,200
Total sheep/lambs	33,131
Total cattle/calves	10,107
Total pigs	4,714

Source: Defra 2008, June Survey of Agriculture and
Horticulture

processing, distribution and retailing) is taken into account; indeed, food processing is the largest manufacturing sector in the UK. The country is characterised by a mixed, diverse and fragmented system of farming on over 300,000 agricultural holdings. In 2008, these holdings accounted for over 17 million hectares of agricultural land, with a majority of this land devoted to permanent grassland and arable crops (see Table 12.1). While the latter are dominated by cereals (especially wheat), the former is the main basis of the UK's livestock industry, involving large numbers of poultry, sheep and cattle.

The UK is nearly 60 per cent self-sufficient in terms of its total food requirements (72.4 per cent for foods that can be grown in the country). However, while levels of self-sufficiency increased from 40–50 per cent in the 1950s to 60–70 per cent in the 1980s, they have been falling in recent years. They also vary significantly between sectors, from very high levels of self-sufficiency in, for example, wheat and potatoes, to between 70 and 85 per cent for the main types of livestock and just 33 per cent for horticulture (fruit, vegetables and salad crops). As a consequence, food imports are increasing with 55 per cent of these coming from the EU. The rest of this section focuses on the fairly self-sufficient sectors of wheat and potatoes and traces the way in which they became modernised and industrialised, as well as integrated into a wider food network that is increasingly dominated by processors and supermarket retailers.

Although the wheat and potato sectors vary significantly in terms of the support they have received under the Common Agricultural Policy (CAP) – a system of EU agricultural subsidies and programmes that initially favoured the production of specific foodstuffs such as cereals, milk and beef – both have seen a trend towards a concentration of production on fewer, larger and more capital-intensive farms. The area devoted to wheat increased continuously until the mid-1980s, since when it has remained at around 2 million hectares; however, output per hectare rose from 6 to 8 tonnes between 1980 and 2000. Today, approximately 30,000 UK farmers grow wheat (compared to 170,000 in the 1950s), with less than half of these accounting for a majority of production. Indeed, the average area devoted to wheat per farm doubled between 1967 and 1997 and is now well over 50 ha, with many farms exceeding this figure by a considerable amount. The registered area of potato production in Great Britain fell from 280,200 ha in 1960 to 124,200 ha in 2007, just as the number of registered potato growers in 2007 was only 3.7 per cent (2,871) of the numbers found in 1960 (76,825). Thus the average area devoted to potatoes by each grower has risen by a massive 1,100 per cent, from just 3.65 ha in 1960 to 43.26 ha in 2007; those growing over 100 ha of potatoes are now producing nearly 40 per cent of total output. Likewise, there has been an almost doubling of yield per ha over the last 50 years, from around 22 to 42 tonnes. This reflects a major increase in the irrigated area of potatoes following the droughts of 1975 and 1976; by 1985, nearly half of the potato crop was irrigated and this has continued to increase ever since.

This process of increasing concentration in wheat and potato production is inherently a spatial one. Use of the location quotient statistic, a ratio measure that compares the area of wheat and potatoes in each county and unitary authority (CUA) with the national distribution of wheat and potatoes – both in relation to the total farmed area in each CUA and Great Britain – helps to show how both crops are now highly spatially concentrated in the more favourable arable farming areas of eastern England including Cambridgeshire, Hertfordshire, Bedfordshire, Lincolnshire and Suffolk (Figure 12.1). Here there is almost continuous cereal cultivation, interrupted only by the planting of break crops such as oilseed rape, sugar beet and peas. As one moves away from this core in western and northern directions, wheat farming decreases significantly. While potatoes used to be grown quite widely throughout Great Britain, they too have become increasingly concentrated in eastern England, with important outliers in Herefordshire, the English–Welsh borders, Shropshire and Lancashire, as well in northeast Scotland – the centre of the seed potato industry. Indeed, there has been a westward and northern shift away from eastern England in recent years in order to find clean land with access to water, and to reduce the risk of disease.

Of the 13.2 million tonnes of wheat produced in the UK in 2007, just over 40 per cent was used for flour milling and just under 40 per cent for animal feed; the remaining 20 per cent was used for exports, distilling and starch. Over 80 per cent of milled wheat is home-grown (around 4.7 million tonnes out of 5.6 million). Not surprisingly, therefore, processors (milling) are the key element in the wheat supply chain or network. A considerable proportion of wheat is marketed through grower cooperatives, who supply wheat to the millers, who in turn sell to various customers including the supermarkets. The trend towards bigger companies characterises the supply chain for

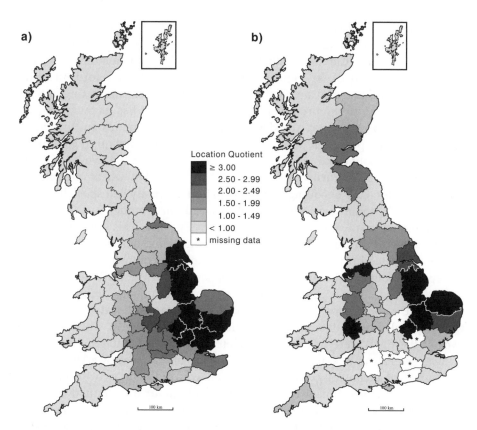

Figure 12.1 The distribution of (a) wheat and (b) potatoes in Great Britain, 2006
Some counties had missing data for total potato production, hence they have no LQ
value.

wheat. Thus there are now fewer agro-chemical companies, seed businesses,
plant breeders, merchants and milling and malting firms. Frontier Agriculture
is one of the key players for UK cereal producers. This vertically-integrated
company provides cereal farmers with many of their inputs, including advice
from agronomists, seed variety trials and fertiliser, and helps them to market
their produce by establishing relationships with the key UK wheat buyers
such as Allied Mills and ABN Feed.

For potatoes, approximately half of total production is sold fresh, compared
to 40 per cent processed (mainly chips and crisps); the remaining 10 per cent
is wasted. The vast majority of fresh potatoes (85 per cent) is sold through the
supermarkets and about 40 per cent of these are pre-packaged. Thus supermarkets are the key cog in this particular supply network, and a decreasing
amount is sold through wholesalers, greengrocers and at the farm gate. The
trend towards larger companies is also apparent, and organisations like
Walkers and McCains are well known for their potato processing activities.
Another important company is Greenvale AP, a leading supplier of fresh

conventional and organic potatoes to retailers, wholesalers, caterers and processors. With three major potato-packing operations in key potato growing areas in Shropshire, Cambridgeshire and Berwickshire, and a number of sales and marketing offices in Norfolk, Suffolk, Somerset, Herefordshire and Yorkshire, it initiates contracts with growers and different marketing outlets. It is also involved in the production of seed potatoes (it has its own branded seed breeding programme), the development of new varieties and irrigation systems, and the provision of advice and technical information to growers.

Government policy and technological change have been the two key drivers of agricultural restructuring in the UK arable sector. For wheat, guaranteed prices under the CAP encouraged farmers to both intensify and specialise their production. Without any upper limits placed on production, the policy favoured larger-scale wheat farming in eastern England. This effect was exaggerated because certain arable crops, notably wheat and oilseed rape, received higher guaranteed prices than livestock products. In contrast, potato farming has operated outside the main support mechanisms of the CAP. Nevertheless, for over 60 years it was controlled by a system of area quotas imposed by the statutory Potatoes Marketing Board (PMB). These were allocated to growers, who could also buy quota area from other registered growers. The key objectives of the PMB were to ensure adequate supplies of home-grown potatoes for consumers and to extend the market for British potatoes (Marsh, 1985).

In terms of technology, wheat production was aided by two 'revolutions'. The first was a mechanical revolution in which larger tractors and combine harvesters helped to transform output per worker and to drive the concentration of wheat production onto fewer and larger holdings. The second was a chemically-based one, where nitrogenous fertilisers, herbicides, pesticides and fungicides helped to increase yields and control weeds, pests and diseases, and encouraged the replacement of mixed farming by intensive systems of specialisation. Similar technological innovations have also been significant in UK potato production. This has involved larger and more expensive machinery, as well as major investment in storage and irrigation facilities – again favouring the larger and more specialised potato producers in eastern England.

Another factor affecting wheat and potato production has been changing consumer demands and, in the last ten years or so, consumers have shown increasing interest in such things as traceability, provenance, branding, local sourcing and wholegrain/healthy food products. This provides an opportunity for organic and exotic varieties, as well as giving some hope to the smaller-scale producer to sell potatoes and wheat-based products directly to the consumer.

Alternative food networks

The UK was hit by a severe outbreak of FMD in 2001. This had a devastating impact on the livestock sector. Thousands of cows, sheep and pigs were slaughtered and burnt in large pyres to prevent the disease spreading. The UK countryside was effectively closed for business during the outbreak. This impacted across all sectors of the rural economy, especially tourism. The disease re-appeared again in 2007, but its spread was less severe. The 2001 FMD 'crisis' marked a significant watershed in public perceptions of UK food networks, and the media attention that came with it highlighted to consumers in quite graphic and emotive ways some of the negative consequences associated with intensive production systems.

The government responded to the crisis by commissioning a report into the FMD outbreak. Published in 2002, the Curry Report was highly critical of UK food supply networks, concluding that food producers and food consumers had become 'disconnected' from one another, that food production had become too intensive and industrialised, and that supermarket chains had become too powerful, influencing not only what and how producers grow food but also what food consumers buy and eat (Holloway, 2008). In making its recommendations, the report called for new ways to 'reconnect' consumers with the food chain and the rural economy. It placed strong emphasis on the potential of farmers to market their produce direct to the end consumer, including strategies to add value to basic commodities (e.g. turning milk into cheese or ice cream, thereby increasing the saleable value of the raw commodity), and the promotion of regional and local foods. The report identified problems in the food chain that geographers and others had noted for some time; for example, the increasing numbers of consumers concerned about how and where their food was produced.

This report thus recognised the potential of 'alternative food networks' (AFNs), especially in a UK context. AFNs are a relatively new and rapidly expanding feature of the UK food economy, typified by the growth in sales of fair trade, organic, local, regional and speciality foods. A significant number of these AFN products can be bought in supermarkets. In the UK, for example, around 75 per cent of organic produce is currently sold this way. For consumers buying AFN products in these retail environments, the connection with the producer remains 'distant', communicated usually via product labels and clever marketing. However, some consumers and producers have established closer relationships by buying and selling food through 'alternative networks' rather than supermarket outlets, including farmers' markets, box schemes and farm shops. These initiatives, which involve building networks that exist outside mainstream food retailing, have grown significantly in recent years (Holloway, 2008).

The growth in farmers' markets has been most spectacular (Kirwan, 2004). These are markets where farmers sell their produce direct to the public. They are exclusive to farmers and usually held in public spaces, with restrictions in terms of how far the producer has travelled (e.g. farm must be within a 50-mile radius of the market). The UK has a rich heritage of 'traditional' retail markets (e.g. in some cases the indoor markets, street markets) where stallholders sell wholesale produce, but farmers' markets are more

Figure 12.2 Organic vegetable box scheme preparing for a delivery, East Sussex (Photo © D. Maye)

recent. The first farmers' market was held in Bath in 1997. Since then the numbers have grown from 200 in 2000 to 422 in England in October 2002 to well over 500 in 2009. These markets have their own geography, with a strong southern concentration but also significant clusters in northern counties such as Yorkshire and Cumbria. Farmers' markets are also popular in Wales, Scotland and, to a lesser extent, Northern Ireland. Vegetable and meat box schemes have experienced similar growth to farmers' markets (see Figure 12.2). From the first one in Devon in 1991, there were over 300 mainly organic box schemes in Britain in 2005, and 2008 figures suggest there are now around 550.

Recent work by human geographers has started to look across different types of AFN to better understand their 'alternativeness'. An excellent example of this is a study by Kneafsey et al. (2008) which involved in-depth case study work with food consumers and producers from six different AFNs: an organic vegetable box, a farmers' market, an urban market garden, a community supported agriculture project (this is an arrangement where people who want to eat the food do not just buy it from the organisation but also become 'subscribers' to the project), a farm shop, and a scheme allowing consumers to 'adopt' a sheep on an Italian farm. This work shows that:

- There is a huge range and number of AFNs operating in the UK, including well-known examples like farmers' markets, as well as more unique, one-off schemes like allotment groups and health and educational projects.

- AFNs are very diverse in terms of how they function and who buys from them. A number of the people interviewed for the study explained how they consume from AFNs for a variety of reasons, and for most they are one part of food shopping, often alongside supermarkets and other 'mainstream' outlets.

- Whilst AFNs are often imagined as a rural phenomenon, a number of highly successful projects were found in urban areas, serving relatively disadvantaged communities as well as affluent consumers.

- AFNs disrupt conventional notions of convenience and food choice as typically articulated by supermarket chains.

- Consumers get pleasure from AFNs – both from the food itself and the social relationships they inspire – and they support relations of care between people and between people and their environments.

Kneafsey et al.'s (2008) work involved research with consumers and producers involved in different AFNs. Other work has been more producer-oriented, examining attempts by producers to develop direct marketing channels such as farm shops, box schemes and farmers' markets. A survey of specialist food businesses in the Scottish–English borders revealed, for example, that a key motivation for producers to engage in these forms of marketing was to better control their supply chain and to retain more of the 'added value' (Ilbery and Maye, 2005). This work also showed how these businesses have to 'dip in' and 'dip out' of conventional food networks because of the way the dominant agri-food system is currently structured (e.g. to source input supplies). Other research on local foods suggests that face-to-face networks provide important social and economic benefits. Morris and Buller (2003) found that farmers in Gloucestershire became involved in the local food sector to re-establish trust with their consumers and to integrate better within the local community. They also found that economic considerations, particularly the higher prices producers can get for products, were another important incentive for farmers.

However, a more recent survey of organic farmers in East and West Sussex by the authors of this chapter suggests that there are some significant difficulties with direct marketing (Ilbery and Maye, 2010). Producers who were originally committed to different forms of direct marketing were now struggling. One of the reasons offered for this was the competition from the large, national 'alternative' forms of direct marketing. The issue of competitiveness was most clearly expressed in the case of box schemes, where local businesses were competing with two large national organic vegetable box schemes (Riverford, and Abel and Cole) who sell to households across the UK. The following quote from an interview conducted in

May 2008 with the owner of one local organic vegetable box scheme in East Sussex summarises well some of the issues:

> 'Boxes have hit the big time, everyone is doing boxes. The milk delivery service, our local greengrocer, the supermarkets, so the concept of boxes is now out there in the market place. Boxes used to be a direct relationship between the people who bought it and the farmer. Now boxes are operated by big, you know, national operators and the supermarkets. The concept has become a product, subject to the whims of the fickle consumer.'

The analysis also revealed a retrenchment away from some forms of direct marketing and a tendency to orientate towards certain types of marketing channel, particularly processors. Similar findings emerged in terms of strategies to add value to primary product (e.g. processing organic milk into organic ice cream). It was often seen as involving much more work. Some organic farmers thus argued that it was often cheaper and more efficient *not* to add value and to sell organic produce directly to marketing co-operatives and/or processors, usually outside the study region. There is also some evidence that organic sales were hit by the economic downturn in 2008–9. A report from the Soil Association (2009) charting trends in 2008 shows, for example, that organic sales to shoppers increased by just 1.7 per cent, after a decade of year-on-year growth of over 20 per cent. However, the report also notes that 'committed' organic consumers were not cutting back. Much of the sales decrease was in supermarkets (Tesco reported a 10 per cent slump, for example), with sales remaining strong in outlets such as farmers' markets.

In this context, AFNs may not offer the rural development benefits promoted by policy-makers. Recently, human geographers have also criticised the localism of AFNs as representing a politics of place which is unreflexive or defensive. In other words, one needs to be sensitive to the dangers of falling into the 'local food trap', wherein 'local' is uncritically accepted as being 'good'. This work also reveals a disjuncture between academic and lay discourses when it comes to AFNs. Surveys of 'alternative' projects have found that practitioners rarely describe themselves as 'alternative' and many in fact resent the term. Recognising these differences, Harris (2009) argues that it may be more constructive to look for challenges *and* opportunities in the diverse practices of AFNs. This is in keeping with the appraisal presented here and all the more important when set against economic cycles, climate change imperatives and the on-going food security crisis.

Food security and the GM debate

The UK farming context has changed quite significantly in the last few years. Much of this is set against a global food economy context (Bridge and Johnson, 2009). Towards the end of 2007, international commodity markets saw sharp price increases for basic food staples – wheat prices doubled, for example, and the price of maize and rice rose by 50 and 20 per cent respectively. Important drivers of this price increase were extreme weather events, notably a severe drought in Australia, and growing consumer demand, particularly in India and China. The Food and Agriculture Organisation (2007) announced that global food reserves were at their lowest level for 25 years and predicted that food prices were likely to remain high in the foreseeable future. Calls were thus made to produce more food on a global level. Whilst food prices have since fallen, as global reserves have re-stabilised somewhat, many argue that 'global food security' is not guaranteed and that it will worsen in the future, with prices likely to remain volatile as energy and water resources becomes increasingly scarce (see Oxfam, 2009). The critical factor driving this new food security dynamic is the world's population, which is predicted to increase from 6 to 9 billion people by 2050. Global food demand is expected to double over the same period, particularly in places like India and China, where it is predicted to increase significantly for meat products.

These changes to the global economy influence UK farming. Since the 1990s, UK farming has become much more aligned to international markets. Market liberalisation has brought price volatility in the past, including very low commodity prices in the early 2000s. However, in recent years economic prospects for UK farming have improved, spurred in part by rising commodity prices. In 2008, total income from farming increased by 9 per cent in real terms, with predictions that this is likely to continue at least until 2011. In this global security context, there is a growing emphasis back towards productivism. This is captured in a speech by Hilary Benn, Secretary of State for Environment, Food and Rural Affairs, to the Oxford Farming Conference in January 2008:

> [O]ne of the greatest challenges we face is ensuring food security in the years ahead. It must be our priority. How will we feed another two and a half billion men, women and children on this small fragile planet of ours in fifty years time, with resources becoming scarce and our climate changing? I want British agriculture to produce as much food as possible. No ifs. No buts. And the only requirement should be, first, that consumers want what is produced and, second, that the way our food is grown both sustains our environment and safeguards our landscape.

Having slipped seemingly out of political fashion, productivism is thus very much centre-stage again in UK farming politics. The critical point is that it has been re-cast in new contexts, with increasing emphasis on producing food in ways that are more competitive, sustainable and resilient. In particular, this involves finding ways to produce food that is less energy dependent. A recent Chatham House (2009) report suggests technology will play a significant role here. This includes re-opening debates over GM technology, also called for in Bridge and Johnson's (2009) *Feeding Britain* report. Some food scientists argue that GM is required in efforts to reconcile demands on agricultural productivity with more sustainable food production. However, it remains a highly contentious issue in the UK, with concerns about health, loss of biodiversity, patenting of crops by large multinationals, and their co-existence with other crops. The Welsh and Scottish governments are both firmly against GM products and have made strong GM-free commitments. This is set to be a significant issue and debates about global food security will continue at UK, EU and global levels.

Conclusion

This chapter has provided an introductory insight into the major aspects of agricultural restructuring in the UK. The established narrative is that agriculture entered a productivist phase from the 1960s to the mid-1980s which was then complemented by a post-productivist phase that persisted until very recently. Various authors have questioned the linearity of this narrative. The examples of cereal and potato farming and AFNs presented here support this critique, revealing an economic geography where different types of food network co-exist and overlap. However, some production networks will be more prominent in some places than others. This is particularly evident with cereal and potato farming, which has continued to specialise and become more geographically concentrated over time. AFNs are more diverse and exist in a variety of places and spaces across the UK. There are some things that make them different to conventional, supermarket-led food networks. Most obvious is the food itself. AFNs place greater emphasis on the quality of the food, in contrast to supermarkets which still tend to favour price and value for money. They also differ in terms of size and scale. Most supply small numbers of consumers compared to supermarkets and the food is often grown on small farms and sold mostly in the local area (Kneafsey et al., 2008). They are thus more locally-embedded.

It is well known that farming makes up an increasingly small part of the rural economy. This can be seen in terms of gross value added (GVA) or employment. In rural England, for example, farming and fishing make up less than 3 per cent of jobs, replaced by activities in service industries and tourism (Ward et al., 2008). Farming is thus being valued in new ways, particularly in terms of its contribution to landscape quality and ecosystem services. More recently, the new 'global food crisis' has arguably placed a re-emphasis back towards productivism. This can be described as 'neo-productivist' in the sense that there is increased emphasis on production, but also with an eye to environmental stewardship and resource, particularly fossil fuel, scarcity. This includes re-opening debates about the potential of biotechnology. It is thus very different to the production mantra initiated in the 1950s. Effects are also likely to be spatially and sectorally differentiated. A cursory survey of UK farm incomes supports this thesis. So, whilst cereal prices have done well thanks to growing global demand, crop shortages in other countries (notably Australia) and the use of some crops for biofuels, livestock producers have done less well, affected by disease outbreaks and low commodity prices relative to rising input costs. This warns against temptations to proclaim a new era of productivism. What seems more likely is that different farming styles and related networks will continue to persist and become more geographically distinct.

Further reading

- Maye et al. (2007) examine debates and practices surrounding efforts to establish 'alternative' systems of food provision.
- Morgan et al. (2006) provide an extensive review of agri-food networks, including so-called 'conventional' and 'alternative' food networks.
- Robinson (2004) reviews geographical aspects of agriculture and food supply systems.

References

Benn, H. (2008) http://www.defra.gov.uk/corporate/about/who/ministers/speeches/hilary-benn/hb090106.htm.

Bridge, J. and Johnson, N. (eds) (2009) *Feeding Britain*. London: The Smith Institute.

Chatham House (2009) *Food Futures: Rethinking UK Strategy*. London: Chatham House.

Curry Report (2002) *Farming and Food: A Sustainable Future*. London: Cabinet Office, Policy Commission on the Future of Farming and Food.

Defra (2008) https://statistics.defra.gov.uk/esg/junesurvey/june_survey.htm.

Food and Agriculture Organisation (FAO) (2007) *The State of Food and Agriculture 2007*. Rome: FAO.

Harris, E. (2009) Neoliberal subjectivities or a politics of the possible? Reading for difference in alternative food networks, *Area*, 41: 55–63.

Holloway, L. (2008) Alternative food networks, *Geography Review*, 22: 10–12.

Ilbery, B. and Bowler, I. (1998) From agricultural productivism to post-productivism, in B. Ilbery (ed.), *The Geography of Rural Change*. Harlow: Longman. pp. 57–84.

Ilbery, B. and Maye, D. (2005) Alternative (shorter) food supply chains and specialist livestock products in the Scottish–English borders, *Environment and Planning A*, 37: 823–44.

Ilbery, B. and Mayes, D. (2010) Clustering and the spatial distribution of organic farming in England and Wales, *Area*, 42, (forthcoming).

Jackson, P., Ward, N. and Russell, P. (2006) Mobilising the commodity chain concept in the politics of food and farming, *Journal of Rural Studies*, 22: 129–41.

Kirwan, J. (2004) Alternative strategies in the UK agro-food system: interrogating the alterity of farmers' markets, *Sociologia Ruralis*, 44: 395–415.

Kneafsey, M., Holloway, L., Cox, R., Dowler, L., Venn, L. and Tuomainen, L. (2008) *Reconnecting Consumers, Producers and Food: Exploring Alternatives*. Oxford: Berg.

Marsh, J. (1985) Economics, politics and potatoes – the changing role of the Potato Marketing Board in Great Britain, *Journal of Agricultural Economics*, 36: 325–43.

Maye, D., Holloway, L. and Kneafsey, M. (eds) (2007) *Alternative Food Geographies: Representation and Practice.* Oxford: Elsevier.

Morgan, K., Marsden, T. and Murdoch, J. (2006) *Worlds of Food: Place, Power and Provenance in the Food Chain.* Oxford: Oxford University Press.

Morris, C. and Buller, H. (2003) The local food sector: a preliminary assessment of its form and impact in Gloucestershire, *British Food Journal*, 105: 559–66.

Oxfam (2009) *A Billion Hungry People.* Oxford: Oxfam International.

Robinson, G. (2004) *Geographies of Agriculture: Globalisation, Restructuring and Sustainability.* Harlow: Pearson.

Soil Association (2009) *Organic Market Report.* Bristol: Soil Association.

Ward, N., Jackson, P., Russell, P. and Wilkinson, K. (2008) Productivism, post-productivism and European reform: the case of sugar, *Sociologia Ruralis*, 48: 118–32.

Wilson, G. (2001) From productivism to post-productivism ... and back again? Exploring the (un)changed natural and mental landscapes of European agriculture, *Transactions of the Institute of British Geographers*, 26: 103–20.

13

THE SHIFTING GEOGRAPHIES OF UK RETAILING

Neil Wrigley

AIMS

- To chart the changing geographies of UK retailing

- To explore how these geographies are shaped in the context of market dominance, tightened regulation and globalisation

- To encourage geographers to examine critically the frequently one-dimensional popular accounts of the transformation of the industry and its consequences

After more than three decades of 'retail revolution' – characterised by progressive consolidation, increasing retail power relative to manufacturers and in the wider economy, and landscapes of consumption transformed by the rise and competitive struggles of major retail corporations (Wrigley and Lowe, 2002) – the UK entered the twenty-first century with several new forces shaping its retail geographies. This chapter focuses on those transformational forces. In particular, it presents a picture of an increasingly concentrated but intensely competitive industry constantly being reshaped by both the evolving regulatory and cultural context in which it operates and by the shifting expectations of consumers regarding the social responsibilities of retailers.

To achieve its objectives the chapter focuses on the food retail sector – the 'lead' sector of the industry which by the mid-2000s accounted for half of total UK retail sales or 13 per cent of total UK household expenditure, and which had a workforce exceeding 1.2 million (5 per cent of UK employees) (Defra, 2006). In particular, the chapter examines how the sector has been transformed by three developments: responses to the emergence of a 'market dominant' lead retailer with visibly growing power; the consequences of a decade of progressively tightening retail planning regulation; and by increasing

Table 13.1 Market shares and contrasting store format distributions of the eight large UK food retailers, 2007

	Share (per cent) of UK grocery sales 2007	Per cent of stores larger than 1,400 sq metres
Tesco	27.6	28
Asda/Wal-Mart	14.1	99
Sainsbury	13.8	53
Morrisons	9.9	96
Somerfield*	3.9	7
Cooperative Group*	3.8	2
M&S Food	3.8	10
Waitrose	3.3	58

Source: Competition Commission, 2008: 32–33
Note: *Somerfield subsequently acquired by Cooperative Group in 2009.

engagement of UK food retailing and its leading firms with the global economy via outward and inward retail foreign direct investment (FDI) and retail-led global sourcing networks.

The emergence of, and responses to, a 'market dominant' lead retailer

Irrespective of the definition of the total UK 'groceries' market which is employed, concentration of the UK food retail industry continued to increase during the early 2000s. By 2007 it had reached a position where eight large chains (see Table 13.1) accounted for 85 per cent of grocery sales totalling £110 billion per annum, and where the four largest firms (Tesco, Asda/Wal-Mart, Sainsbury and Morrisons) alone accounted for just over 65 per cent. In particular, the market share of the lead retailer (Tesco) accelerated during this period – increasing by over one third – to a level of between 26 per cent and 31 per cent depending on the definition of the total market used. Of itself, this represented a potential 'scale monopoly' (defined as when more than a quarter of goods of a particular type are supplied in the UK by the same firm) and many commentators – both academic (Burt and Sparks, 2003) and popular (Simms, 2007) – viewed it as the emergence of 'market dominance'. In turn, it provoked an extended period of attempts to 'rein in' that growing dominance of the market leader and also to address wider distortion-of-competition issues arising from the highly consolidated nature of the industry. Those attempts were driven by environmentally focused pressure groups and non-governmental organisations (NGOs) (e.g. Friends of the Earth, Campaign to Protect Rural England), and they sometimes involved orchestrated campaigns (e.g. the 'Tescopoly' campaign of the New Economics Foundation). However, they also reflected sustained small-trader induced Parliamentary pressure to instigate regulatory action which was placed on the Office of Fair Trading and Competition Commission (e.g. pressure to investigate the concerns expressed in the *High Street Britain 2015* report of the All-Party Parliamentary Small Shops Group). Taken together, these attempts and their consequences have provided one of the main motifs of the shifting geographies of UK retailing in the early part of the twenty-first century.

Over a period of ten years, beginning with the major *Supermarkets inquiry* (Competition Commission, 2000) of 1998–2000 and ending with the comprehensive *Market Investigation into the Supply of Groceries* of 2006–08 (Competition Commission, 2008), four major investigations were conducted by the UK regulatory authorities into competitive conditions in the food retail industry. It is reasonable to summarise this period of intense regulatory scrutiny as producing mutual misunderstanding and frustration for all parties. For the NGOs and small-trader organisations it was axiomatic that the increasing consolidation of the UK food retail industry had resulted in profoundly negative and anti-competitive consequences – ranging from intolerable pressures being placed on both UK and international suppliers, through the potential destruction of the UK small and specialist retailer sector with an accompanying loss of diversity and amenity, to the homogenisation of UK high streets and the creation of what the New Economics Foundation (2005) described as 'clone town Britain'. From the perspective of the major corporate retailers, however, it was unquestionable that they operated in an intensely competitive industry, and that their success merely reflected the preferences of, and implied a tacit contract with, the consumers they served. In general, they viewed this period of sustained regulatory pressure as unnecessary and unjustified given their significance to the UK economy, their important role in reducing consumer price inflation, and their consumer mandate. Finally, from the perspective of the regulatory authorities, the issues were strictly those of ensuring that markets (particularly 'local' markets) remained competitive, open to entry, and that consumer welfare was protected. Also, that the practices employed by the retailers – ranging from the possible imposition of barriers both to market entry and exit, through potentially using their buying power to exert anti-competitive supply chain conditions, to distorting competition via their pricing policies – were not adversely affecting competition or harming the consumer. The Competition Commission's (2007: 34) position, therefore, was that it should not attempt 'to impose on the consumer [its] own judgement of what the grocery retailing offer should be'.

In summary, the findings of these investigations expressed:

- increasing concern about the competitive structure of local markets in certain areas of the UK – that is to say, areas where the leading food retailers enjoyed effective 'monopoly' or 'duopoly' status;

- growing determination by the regulatory authorities to prevent that situation deteriorating as a result of merger or acquisition activity – essentially by severely restricting such activity if it involved the leading firms (e.g. when Safeway was permitted only to be acquired by Morrisons in 2003), and by requiring permitted acquisitions to proceed only with substantial divestment of 'monopoly/duopoly' stores (e.g. the post-acquisition divestment required of ex-Safeway stores by Morrisons in 2003–4, of ex-Morrisons stores by Somerfield 2005–6, and ex-Somerfield stores by the Cooperative Group in 2009);

- a growing resolve to align competition and retail-planning regulation – ultimately in the form of a proposed 'Competition Test' (Competition Commission, 2008) – that is to say, by moves to reverse the previous 'competition blind' position of UK planning regulation where the identity of the firm proposing a retail development was not a factor taken into account by the planning authorities; and

- belief in the value of implementing (and progressively strengthening) a Code of Practice relating to the supply-chain practices of the leading retailers.

Despite these concerns and resolutions, however, the overall judgement of these prolonged and extensive investigations of the UK food retail industry by the regulatory authorities remained consistently that:

the industry is currently broadly competitive and, overall, excessive prices are not being charged nor excessive profits earned (Competition Commission, 2000, Vol. 1: 7)

and that:

in many important respects, competition in the UK groceries industry is effective and delivers good outcomes for consumers. (Competition Commission, 2008: 9)

More specifically, in the context of concerns surrounding the emergence of a 'market dominant' lead firm (Tesco), and the impacts of both that firm and other leading food retailers on the small-trader independent convenience store sector, the Commission (2008: 9–10) essentially ruled that those views were not substantiated.

The NGOs, pressure groups and small-trader organisations found the judgements to be extremely disappointing – a product as they saw it of an over-tightly drawn 'economistic' view of the issues taken by the Office of Fair Trading (OFT) and the Competition Commission. Nevertheless, some key aspects of the case they had advanced to initiate the inquiries were found not to hold up to the scrutiny. In particular, claims made by the Association of Convenience Stores (ACS) that independent convenience stores had suffered a sharp decline in the early 2000s as a result of the entry of Tesco and Sainsbury into the market were dismissed by the Commission, which concluded that it did *not* consider independent convenience stores to be in broad decline, and additionally did *not* believe that British high streets were experiencing an *accelerating* decline in their small and specialist stores.

Instead it provided counter-evidence of considerable new entry and robust growth in independent convenience store numbers during the early 2000s in parts of the UK – evidence confirmed by a parallel study of over 1,000 UK town centres and high streets (Wrigley et al., 2009) which reported large increases in such stores, particularly in London and the South-east, in part resulting from the so-called 'Polish grocer' effect (see Chapter 17). Additionally, and even more controversially, the Commission's analysis suggested that, set within a general context of continuing long-term decline of specialist small stores in UK town centres and high streets, competitive corporate supermarket entry was not inevitably and uniformly associated with negative impacts on the small store sector. Again, these findings were confirmed and extended in the parallel study (Wrigley et al., 2009) which showed that in the 'growth Britain' region of 'London and Prospering Southern England' many types of small and specialist stores performed more robustly in the face of competitive supermarket entry than was commonly believed.

Nevertheless, despite what the campaigning NGOs and small-trader organisations saw as a failure of the regulatory authorities to 'rein in' the power of the major food retailers, and the 'market dominant' lead retailer in particular, it is clear that the sustained media pressure which they exerted during the early to mid-2000s did penetrate deeply into the public psyche and had a material effect on the UK food retail industry and its geographies. Indeed, as a report on the industry by the Department of Environment and Rural Affairs (Defra) made clear, success of the leading UK food retailers had:

> historically been predicated on high levels of public and political support, but that is no longer guaranteed. Concerns and adverse publicity, well-founded or not, reflect growing expectations regarding the social responsibility of supermarkets. (Defra, 2006: 23)

What is equally clear, however, is that these major retailers (in particular, the agile lead firm) had responded during the early 2000s in what Defra described as:

> innovative, dynamic and generally profitable ways to rising consumer expectations, within the evolving regulatory and cultural framework in which they operate[d]. (2006: 23)

In particular – not least via their prioritisation of community responsiveness, adoption of urban regeneration agendas, and commitments to sustainable development and ethical/responsible sourcing – the stereotype corporate retail targets of the campaigning groups responded by stealing the arguments of their critics. Nowhere was this more obvious than in the adaptation of the retailers' store development programmes to a decade of tightening retail planning regulation – the issue to which the chapter now turns.

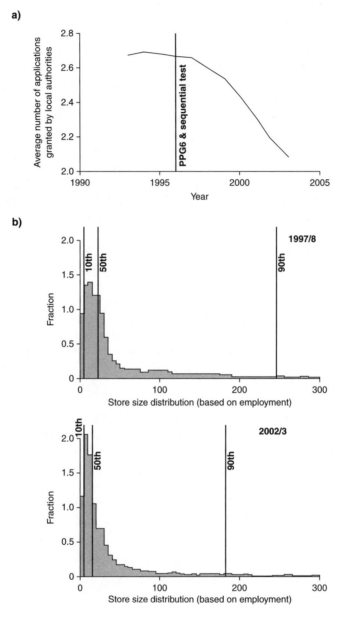

a) reports number of major retail applications granted across 304 English local authorities 1993–2003

b) shows changing average size of non-specialised stores (mostly supermarkets) operating nationally (in all 11 UK regions) as measured by employment in each store. Vertical lines mark 10th, 50th and 90th percentiles of distribution

Figure 13.1   After 1996, obtaining planning permission for large retail stores became more difficult (a), and large food retailers began opening smaller stores (b) (Redrawn from Sadun, 2008)

The consequences of progressively tightening retail planning regulation

Beginning in earnest in the mid-1990s with the 1996 revision of *Planning Policy Guidance (PPG) Note 6* and its incorporation of the so-called 'sequential' test of new retail development proposals, retail planning policy in the UK moved into a decade-long phase of progressive tightening (Guy, 2006). The viability and vitality of town centres was prioritised and the sequential test required development of town-centre sites, followed if necessary by edge-of-centre and district-centre sites and, only as a last resort, out-of-centre sites. Additionally, following a High Court judgement in 1998, developers of proposed out-of-centre sites were required to demonstrate that there was a 'need' for the retail space. As a result, obtaining planning permission for the types of large-format out-of-centre retail developments which had characterised the 1980s and early 1990s became significantly more difficult (Figure 13.1(a)).

By the start of the 2000s the 'town centres first' policy had become increasingly effective. Ministerial statements and responses to Parliamentary Committee reports during the late 1990s had effectively tightened the application of the sequential test and introduced a stricter 'class of goods' interpretation (could the goods to be sold in a proposed out-of-centre development be sold instead from town-centre sites), and planning appeal decisions tended to confirm that stricter interpretation. In addition, broad cross-political-party support continued to underpin the policy. Retail development had not disappeared – rather, it had adjusted to the new realities. The major food retailers (particularly Tesco) had adapted to the constraints of the new era via increased flexibility regarding both store formats and the sites they were potentially willing to develop. One consequence was that the average size of the new stores opened by those firms began to fall in the late 1990s as they first gained experience of, and subsequently began to roll-out, smaller 'planning-regulation friendly' formats more suited to constrained town–centre or edge-of–centre sites. Wrigley (1998) details the early phases of this shift in terms of the rising proportion of smaller-format 'Compact', 'Metro' and 'Express' stores contained within Tesco's development programme in the mid 1990s, whilst Figure 13.1b illustrates declining store sizes between 1997–98 and 2002–3. Another consequence was that the major retailers began to reassess more generally the potential of what, for most of the previous 20 years, had been regarded as marginal locations of food retail profit extraction. That reassessment began with town centres and high streets which since the 1980s had been losing their major-retailer food stores to out-of-centre developments, and subsequently widened to include 'brownfield' and deprived Local Authority housing estate sites. In other words, as planning regulation progressively tightened, locations which for two decades had been largely passed over for new store development appeared to offer increasingly attractive potential.

As they adapted their store development programmes to work *with the grain* of the 'town centres first' policy, the major retailers were not adverse to exploring and exploiting flaws in planning legislation in order to maximise the yield from their existing stores. Indeed, progressively tightening planning constraints provided them with strong motivation to seek ways of using their existing sites more intensively. The two clearest examples of this were: first, the significant shift of investment into store extensions in the late 1990s/early 2000s (Wrigley, 1998) until extensions were incorporated within the remit of the sequential test in 2003, and second, the active exploitation between 2002 and 2005 of the so-called 'mezzanine floor planning loophole' (Wood et al., 2006) until extension of the size of stores *within* their existing footprint in that way was rapidly blocked by revised legislation in 2005. More generally, however, adaptation via working *with the grain* of planning policy had two important implications. First, it altered the organisational range and operational skills of some of the major food retailers, with Tesco and Sainsbury (but not Asda/Wal-Mart or Morrisons – see Table 13.1) becoming increasingly 'multiformat' operators. Second, it reflected a growing realisation by those firms that their corporate growth prospects were intrinsically intertwined with evolving public expectations regarding their social and community responsibilities. In what ways then have these issues been reflected in the retailers' store expansion programmes during the first decade of the twenty-first century? Three examples usefully illustrate these themes.

The first relates to the adoption of urban regeneration agendas by the major retailers and the development of significant numbers of 'urban regeneration partnership stores' in some of the most deprived urban communities in the UK (Wood et al., 2006). Partnerships involving the major retailers, local authorities, labour unions, government employment agencies and community organisations sought to kick-start regeneration in deprived communities with the help of private sector investment and to link regeneration with programmes of skills training and employment favouring the local community. While these schemes were initially castigated by some commentators as merely clever devices to get stores built and passed by planners during a period of tightening retail planning regulation, and whilst the retailers freely acknowledged that their involvement related to both obtaining sites and securing a workforce in the tight labour market conditions of the early 2000s, they also argued that their motivation for involvement was best described as *'enlightened* self interest'. In other words, acceptance by the major retailers that their corporate growth prospects were intrinsically bound up with evolving public expectations regarding their social and community responsibilities resulted in a growing willingness to serve socially excluded deprived urban communities, thereby reversing development trends which over the previous 20 years had been responsible for the creation of the so-called 'food deserts' that had captured the imagination of policy makers in the late 1990s (Wrigley, 2002). The fact that those underserved communities offered both a feasible expansion trajectory for large-format stores when other possibilities had progressively been shut off, together with the possibility of tapping into potentially significant pockets of under-exploited demand, does not deflect from the strong case the retailers were able to mount for involvement in such schemes based around the contribution (not least the targeted employment benefits) they could make to tackling social exclusion in the UK. Indeed, these claims were essentially accepted in the revision of PPG6 which came into force in 2005. That revision, referred to as *Planning Policy Statement (PPS) 6*, incorporated regeneration of deprived communities and promotion of social inclusion as relevant matters which could be treated as 'material' in obtaining planning permission for retail development.

The second example concerns the rapid expansion of some of the major retailers (especially Tesco and Sainsbury, but also the Cooperative Group) into the convenience store sector since 2002. Some commentators have interpreted this expansion as being essentially the result of the so-called 'two market' ruling of the Competition Commission which viewed the 'one stop' grocery shopping market in the UK (stores above 1,400 square metres) as a separate sector from the 'secondary' grocery shopping market (stores below that size). That ruling gave the regulatory 'green light' for Tesco's acquisition of the 862-strong T&S chain in 2002, followed by the Adminstore chain in 2004, and the subsequent conversion of many of those stores into the Tesco 'Express' format, together with Sainsbury's acquisition of the Jackson, Bells and Beaumonts chains in 2004 (Wood et al., 2006; Wrigley et al., 2009). However, to describe the move of these retailers into the convenience store sector as being merely an opportunistic response to a competition regulation loophole would be to significantly underplay the consequences of their adaptation to working *with the grain* of planning policy. As a result of increasing the flexibility of formats and being forced to master the organisational and operational skills of smaller-format retailing during the period of regulatory tightening, these firms had placed themselves into a position where they could operate networks of smaller stores with levels of efficiency and profitability approaching those that had characterised larger-format stores in the 1980s. Moreover, being profoundly consumer-facing and consumer-driven organisations, they had detected the growth during that period of what might be termed 'convenience culture' – a rising demand for 'neighbourhood' stores (involving less travel and shopping time commitment than large out-of-centre stores) but which nevertheless still offered (as a result of efficient logistic systems) availability of the high-quality fresh produce associated with the superstores. That is to say, a 'choice edited' version of the larger-format stores. The consequence was that, despite intense campaigning pressure by small-trader organisations and NGOs against the roll-out of the networks of corporate convenience stores, the essential responsiveness of those retailers to evolving public expectations resulted in far greater community support for the corporate convenience stores than the campaigning groups might have expected.

The third example concerns the progressive and increasingly rapid shift of the major retailers into environmentally sustainable store development. By 2009, after several years of experimenting with increasingly 'environmentally friendly' architectural designs and engineering for a new generation of 'low-carbon' stores, the lead retailer was able to announce the opening of a store in Manchester which had a carbon footprint 70 per cent smaller than any of its previous stores, with plans for zero-carbon developments at an advanced stage. Whilst in some ways these might be regarded as merely the product of the retailers' relentless drive for cost efficiencies, and as a response to increasingly stringent government directives on emissions and the threat of future 'green taxes', the retailers themselves have again taken an *'enlightened self-interest'* view of this shift. Once again, this illustrates their continuous sensitivity to both shifting consumer expectations and emerging regulatory agendas – in particular in this case to the possible emergence of a new phase of UK retail planning regulation which will place emphasis on the delivery of sustainable development in both economic and environmental terms. As that policy emerges – beginning during 2009 with the launch of the integrated policy statement (PPS4) *Planning for Sustainable Economic Growth* – and is actively shaped by the retailers, the leading firms can be expected to respond and adapt to it with levels of dynamism and innovation similar to those characterising their response to the decade-long period of regulation tightening.

The consequences of engagement with the global economy

Finally, and intrinsically interrelated to the previous two themes, what have been the consequences during the early years of the twenty-first century of the progressively increasing engagement of the UK food retail industry and its leading firms with the global economy? Three aspects of that engagement deserve attention.

The first relates to outward retail FDI from the UK, and the relative levels of growth potentially available during the late 1990s and 2000s from operating in emerging-market economies with less organised 'modern' retail sectors as compared to operating in 'mature' markets such as the UK. In this context, as repeated attempts were made to 'rein in' the growing dominance of the market leader via pressure on the regulatory authorities to investigate competitive conditions in the industry, as tightening planning regulation slowed and modified the nature of store expansion programmes in the UK, and pressures to link tightening retail planning regulation to the competitive structure of local markets began to mount, Tesco responded by shifting at an ever increasing rate the main drivers of its longer-term expansion into those potentially higher-growth-rate international markets. The resulting transformation of Tesco was remarkable. In 1997 it had less than 8 per cent of its operating space outside the UK and no presence in Asia. By 2008 it had approximately 60 per cent of its operating space outside the UK, had a significant presence in 12 international markets, and had over 800 stores accounting for almost 30 per cent of its operating space in Asia (Figure 13.2 illustrates its impact in Thailand). Over that period Tesco's outward retail FDI amounted to over £10 billion, and it had essentially reinvested the 'free cash flow' from its UK core business into building an international 'engine of growth' which could sustain the firm's continued expansion during the twenty-first century. During

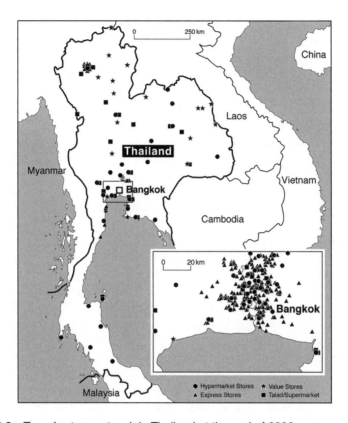

Figure 13.2   Tesco's store network in Thailand at the end of 2006

Source: author's research

that transformation from purely domestic-market operator to leading multinational – becoming the world's third largest retailer by 2008–9 – it was involved in a process of *continuous* organisational change. Not least, this resulted from learning to operate in, and transferring the knowledge obtained from, host economies/societies with very different institutional, cultural and regulatory characteristics (Coe and Wrigley, 2007). The consequence, however, was that by the end of the first decade of the twenty-first century, UK retailing had a lead firm which had a profoundly different level of engagement with the global economy than had been the case a decade earlier. Its overall performance and prospects were significantly less dependent on the UK economic cycle than had been the case in the mid-1990s and, by operating across multiple regions and markets with contrasting business systems, its resilience to the shifting trajectories of the global economy had been fundamentally altered. It was simultaneously more exposed to the risks of the global economy (in the sense of an unexpected political/economic crisis impacting one of its international subsidiaries), whilst at the same time able to benefit from those risks by being able to find and/or maintain growth when conditions for core-market growth were difficult.

The second and inverse aspect of global economy engagement relates to inward retail FDI into the UK. During the late 1990s and 2000s the competitive landscape of the UK food retail industry was transformed by the entry and expansion of international retailers. There were two dimensions of that inward investment – entry, via acquisition, of the world's largest retailer, Wal-Mart, in direct competition with the incumbent major operators and *de nouveau* entry followed by organic expansion of a group of European limited-assortment hard discount retailers (LADs). In the case of the LADs, following entry into the UK market in the early 1990s and initial slow expansion, Aldi, Netto and Lidl began to develop substantial store networks. Focussing on a market niche which had been neglected by the major UK food retailers, the LADs typically targeted lower-income urban communities, selling a limited range of products at very competitive prices from smaller-format stores averaging 8,500 square feet. Unconstrained by the competition regulation which increasingly impacted the major 'one stop' food retailers, and developing sites which were more likely to meet planning regulation requirements, their store opening programmes and market share began to accelerate during the 2000s. By 2007, the LADs operated a combined 1,020 stores (Lidl having 465), generated sales of £3.1 billion per annum, almost 5 per cent of the total UK groceries market (Competition Commission, 2008), and were significantly altering the landscape of urban food retail provision in the UK. Moreover, as financial crisis and global recession took hold during 2008–09, the hard discount offer of the LADs, tailored to the price-sensitive consumer, prompted both a marked increase in their sales and a ramping-up of the pace of their store expansion programmes. By contrast, Wal-Mart's entry into the UK, via a £6.7 billion acquisition of Asda in 1999, was more dramatic, initially promising to alter competitive conditions in the market in a fundamental way (Burt and Sparks, 2001) as a result of Wal-Mart's global buying scale and sourcing networks. In practice, however, Wal-Mart was unable to realise that potential in a regulatory environment in which it was effectively blocked from achieving market leadership via additional acquisitions, and in which it proved less agile than Tesco in expanding its store base via adaptive change to format and site selection. Indeed, for much of the 2000s, Asda/Wal-Mart's share of the UK groceries market remained static at between 14 per cent and 16 per cent, depending on the definition of the market used, whilst Tesco's share increased significantly. Nevertheless, Wal-Mart's market entry did more gradually begin to have a material impact on the UK regulatory environment. In particular, as its frustration with UK planning and competition regulation progressively increased, its global scale gave it the political clout to lobby government (particularly the Treasury) for a closer examination of the UK retail industry's relatively poor productivity performance, and to nudge the Treasury towards a more 'growth' oriented policy position. That view – reflected in Treasury statements and in the *Barker Review of Land Use Planning* (2006) – became an important counterweight to the more interventionist regulatory position of the government departments responsible for planning.

The third aspect of global economy engagement relates to the increasing involvement of the major UK food retailers in retail-led global sourcing networks. Perhaps surprisingly, given Gereffi's (1994) highlighting of the role of retailers as lead firms in 'buyer-driven' global commodity chains, that involvement did not become as pervasive as many commentators would believe until the late-1990s. However, from that point onwards, as the work of Dolan and Humphrey (2000) and others clearly demonstrates, the close management of global and regional supply chains, which become increasingly 'demand pull' in character, was an important aspect of the geographies of UK food retailing. In particular, as UK retailers increasingly co-ordinated global sourcing networks and progressively imposed their 'private standards' of quality and safety on suppliers, by default they became deeply implicated in the governance of those networks. Evolving consumer expectations regarding the social responsibility of the supermarkets manifested itself in demands that the major food retailers exercised 'care at a distance' within their sourcing networks – not least in respect to labour conditions at sites of production. In turn, this was reinforced by the increasing significance placed by the investment community on the ethical performance of large corporations. The consequence was that the UK food retailers increasingly came to understand the value of, and proactively adopted, ethical trading/responsible sourcing agendas – most notably via the UK-based multi-stakeholder Ethical Trading Initiative (ETI) (see Hughes, 2005). And, in turn, that was merely a part of a wider adoption of corporate social responsibility (CSR) agendas by those firms. Indeed, in the case of Tesco with both extensive global sourcing networks *and* store networks in many international markets, an important issue became one of how to manage, via its CSR programmes, the 'global responsibilities' associated with being an increasingly major multinational organisation.

Conclusion

This chapter has outlined some of the complex interactions between an increasingly concentrated but intensely competitive industry, ongoing regulatory re-interpretation, and progressively deepening engagement with the global economy, which together have shaped the shifting geographies of UK food retailing during the first decade of the twenty-first century. During that period, consumer expectations regarding the social responsibilities of the major food retailers have evolved – not least in response to the sustained pressures placed on the major firms in the industry by the campaigning groups. Often seemingly in anticipation of those changing expectations, reflecting closeness to their customers, the major retailers (particularly the lead firm) have quickly embraced the agendas of community responsiveness, urban regeneration, sustainable development and ethical/responsible sourcing.

The view presented in this chapter is that the outcomes of these complex interactions have never and will never be straightforward to read off. On the one hand, the major retailers are clearly locked in a process of continuous dialogue with government which shapes the outcomes. On the other hand, as is well illustrated by Tesco's transformation from predominantly domestic operator to multinational giant, those retailers do not remain static as organisations. It is the combination then of adaptive and agile responses to regulatory re-interpretation, sensitivity to evolving consumer expectations of their social responsibilities, the development via dialogue of initiatives which coalesce with government agendas, and continuous organisational transformation of the major retailers – not least as a result of engagement with the global economy – which must be understood to develop an appreciation of the shifting geographies of UK retailing. The processes are complex and subtle. Their interpretation demands that geographers critically examine the limitations of the frequently uni-dimensional popular accounts of the transformation of the UK retail industry over recent decades, and that they attempt to understand the multi-dimensional and conflicting perspectives (often simultaneously held) of government, regulators, retailers, suppliers and consumers.

Further reading

- Wrigley (2009) offers an up-to-date introduction to key themes in retail geography, particularly the economic themes underpinning this chapter.
- Burt and Sparks (2003) provide a clear account of the emergence and causes of market dominance in the UK food retail industry.
- Wood et al. (2006) give a comprehensive summary of adaptive responses of major retailers to decade of tightening planning regulation.

References

All-Party Parliamentary Small Shops Group (2006) *High Street Britain, 2015*. www.nfsp. org.uk.

Barker Review of Land Use Planning (2006) HM Treasury, London. Available from: http://www.hm-treasury.gov.uk/barkerreview_land_use_planning_index.htm

Burt, S.L. and Sparks L. (2001) The implications of Wal-Mart's take over of Asda, *Environment and Planning A*, 33: 1463–87.

Burt, S.L. and Sparks, L. (2003) Power and competition in the UK retail grocery market, *British Journal of Management*, 14: 237–54.

Coe, N.M. and Wrigley, N. (2007) Host economy impacts of transnational retail: the research agenda, *Journal of Economic Geography*, 7: 341–71.

Competition Commission (2000) *Supermarkets: A Report on the Supply of Groceries from Multiple Stores in the United Kingdom*, Cm4842, October.

Competition Commission (2007) *Market Investigation into the Supply of Groceries in the UK: Provisional Findings Report*, October.

Competition Commission (2008) *Market Investigation into the Supply of Groceries in the UK: Final Report*, April.

Department of the Environment (1996) *PPG6 (revised): Town Centres and Retail Development.* London: HMSO.

Defra (2006) *Economic Note on UK Grocery Retailing.* London: Department for Environment, Food & Rural Affairs, Food & Drink Economics Branch.

Dolan, C. and Humphrey, J. (2000) Governance and trade in fresh vegetables: the impact of UK supermarkets on the African horticultural industry, *Journal of Development Studies,* 37: 147–76.

Gereffi, G. (1994) The organization of buyer-driven commodity chains, in G. Gereffi and M. Korzeniewicz (eds), *Commodity Chains and Global Capitalism.* Westport, CT: Praeger. pp. 95–122.

Guy, C.M. (2006) *Planning for Retail Development.* London: Routledge.

Hughes, A.L. (2005) Corporate strategies and the management of ethical trade: the case of UK food and clothing retailers, *Environment and Planning A,* 37: 1145–63.

New Economics Foundation (2005) *Clone Town Britain,* June. London.

Office of the Deputy Prime Minister (ODPM) (2005) *Planning Policy Statement 6 (PPS6): Planning for Town Centres.* London: ODPM.

Planning Policy Statement 4 (PPS4): *Planning for Sustainable Economic Growth.* Department for Communities and Local Government (DCLG). London, 29 December 2009.

Sadun, R. (2008) Supermarkets and the British high street: the unintended consequences of planning regulations, *Centre Piece,* 13, 2: 2–5. London School of Economics: Centre for Economic Performance.

Simms, A. (2007) *Tescopoly: How One Shop Came Out on Top and Why it Matters.* London: Constable and Robinson.

Wood, S.M., Lowe, M.L. and Wrigley, N. (2006) Life after PPG6 – recent UK food retailer responses to planning regulation tightening, *International Review of Retail, Distribution & Consumer Research,* 16: 23–41.

Wrigley, N. (1998) PPG6 and the contemporary UK food store development dynamic, *British Food Journal,* 100: 154–61.

Wrigley, N. (2002) Food deserts in British cities: policy context and research priorities, *Urban Studies,* 29: 2029–40.

Wrigley, N. (2009) Retail geographies, in R. Kitchen and N. Thrift (eds), *International Encyclopaedia of Human Geography,* Vol. 9. Oxford: Elsevier PP. 398–405.

Wrigley, N. and Lowe, M.S. (2002) *Reading Retail: A Geographical Perspective on Retailing and Consumption Spaces.* London: Arnold.

Wrigley, N., Branson, J., Murdock, A. and Clarke, G.P. (2009) Extending the Competition Commission's findings on entry and exit of small shops in British high streets: implications for competition and planning policy, *Environment and Planning A,* 41: 2063–85.

14

UK ENERGY DILEMMAS: ENERGY SECURITY AND CLIMATE CHANGE

Michael J. Bradshaw

AIMS

- To review recent trends in UK energy production and consumption

- To assess the energy security challenges that now face the UK

- To consider the consequences of the green house gas emission targets set by the UK government

---

The UK Government's White Paper on Energy (DTI, 2007: 6), *Meeting the Energy Challenge*, noted that 'energy is essential in almost every aspect of our lives and for the success of the economy'. It then identified two long-term energy challenges: first, 'tackling climate change by reducing CO_2 emissions both within the UK and abroad'; and second, 'ensuring secure, clean and affordable energy as we become increasingly dependent on imported fuel'. This is the crux of the UK's energy dilemmas: seeking to orchestrate a transition to an affordable low-carbon energy system and economy while being increasingly dependent upon external sources of energy supply.

The energy sector has been neglected in recent research on the economic geography of the UK. In the past, geographers at Durham studied the demise of the coal industry in the north-east of England (Beynon et al., 1991), while others in Scotland studied the development impact of North Sea oil and gas (Cumbers et al., 2003). This more recent neglect may be because de-industrialisation and the growth of the service economy have led us to forget the material basis of the economy (see Bridge, 2009). The smoke-stack industries may have gone, but energy services remain essential to the UK economy. Equally, the abundance of domestic supplies of oil and gas has made UK policy makers complacent about energy security. The harsh reality is that North Sea oil and gas production is declining and much of the UK's existing energy infrastructure is ageing and must be replaced in the near future. Therefore, when oil prices peaked at \$147 a barrel in July 2008 and when, in January 2009, a dispute between Russia and Ukraine resulted in shortages of natural gas in Europe (a repeat of the 2006 dispute, but with more lasting consequences), the UK's reliance on cheap energy and its increasing vulnerability to supply disruptions became all too apparent. At the same time, international consensus has finally emerged regarding the threat posed by climate change. In the UK, the energy sector accounts for 95 per cent of CO_2 emissions (DEEC, 2009a); thus policies aimed at achieving substantial reductions in CO_2 emissions must consider the energy sector.

This chapter provides an introduction to the energy challenges that now face the UK if it is to succeed in making the transition to a low carbon economy. The scale of these challenges should not be underestimated; in historical context they represent a change in the relationship between energy, economy and environment akin to the industrial revolution. This chapter is divided into three substantive sections. The first section provides some essential historical background and explains the fundamental changes that have taken place in the political economy of the UK's energy system since the end of the Second World War. The second section considers the energy security challenges that now confront the UK. The third section brings climate change into the mix and examines trends in greenhouse gas (GHG) emissions and considers the consequences for the energy economy of the targets set by the UK government for reductions in GHG emissions.

A brief recent history of UK energy

Today, in the UK, most people take it for granted that when they turn on the light switch, the light goes on. Seldom do we experience power cuts, and if we do, they are usually short-lived. But, consider the fact that, according to the United Nations, worldwide 2.4 billion people still rely on traditional biomass (wood, animal dung etc.) for cooking and that 1.6 billion people do not have access to electricity. A secure and affordable supply of energy (a simple definition of energy security) is something we consider a public good. It is for this reason that energy security is the business of the state, even if, in the UK at least, the energy business is in private hands.

The current UK energy system is the result of the policies of privatisation and liberalisation orchestrated by the Conservative government of Margaret Thatcher after 1979 and then continued by subsequent Labour governments. In 1980 the state owned gas, coal, nuclear and electricity industries as well as having a major stake in the oil industry. By 2000 it had sold off all but a share in the nuclear industry, and today it owns nothing; instead, the market and the private sector is now the guarantor of secure and affordable supplies of energy. This was not always the case in the UK and is still not the case in many other EU member states. In the UK up until the 1980s it was conventional wisdom that markets were hopelessly inadequate at providing appropriate energy supplies. Helm (2003: 3) describes the period between the end of the Second World War and 1980s when: 'the overwhelming objective was to produce as much energy as possible (domestically) to keep pace with the demand of what has become known as the "golden age" of the British economy.' For most of the post-war period the UK's energy sector was developed by the state through integrated monopolies (see Table 14.1), and much of today's infrastructure was built during this period. As a consequence, when privatised, the new owners inherited assets, price levels and structures that were not designed to maximise efficiency, but rather to meet wider social and political objectives. It is not the purpose of this chapter to delve in any detail into the causes and consequences of the privatisation of the UK's energy sector that took place in the 1980s (see Helm, 2003, for that). The key point is that the current system is based on an infrastructure that, for the most part, was built under an entirely different logic.

Table 14.1 Ownership changes in the UK energy industry

Sector	Pre-war ownership	Nationalised	Privatised[*]
Coal	Private	National Coal Board (1947)	RJB Mining and others (1995)
Electricity	Central Electricity Board, municipalities and private companies	British Energy Authority (1948) and the Central Electricity Generating Board Areas Board, and the Electricity Council (1957)	National Power, Power Gen (1990) Regional Electricity Companies (1990) Scottish Power and Scottish Hydro-Electric (1991)
Gas	Municipalities and private gas undertakings	Area Boards and the Gas Council (1948), and then British Gas Corporation	British Gas (Gas Act 1986)
Oil	Anglo-Iranian Oil Company	BP (partial), British National Oil Company (1977)	BP final sale (1987) Britoil (1982) Enterprise Oil (1984) British Energy (1996)
Nuclear	None	United Kingdom Atomic Energy Authority (1954), British Nuclear Fuels (1971) Nuclear Electric (1990) Scottish Nuclear (1990)	British Energy (1996)

Source: Helm, 2003: 18
Note: * Dates refer to vesting of assets, except BP.

The fundamental changes that have restructured both the pattern of ownership and energy production and consumption in the UK have their origins in the early 1970s, when the Arab members of the Organisation of Petroleum Exporting Countries (OPEC) cut oil supplies to protest Western support for Israel in the 1973 Arab–Israeli war. The OPEC embargo resulted in a step change in the price of oil which prompted industrial economies, such as the UK, to promote energy efficiency and reduce energy-intensive industries. The result was a sustained improvement in energy intensity, with less energy being required year-on-year to generate a unit of economic output. According to the International Energy Agency (IEA, 2007: 65), between 1973 and 2004 the total amount of primary energy supply (TPES) per unit of gross domestic product (GDP) fell in the UK by 46 per cent, compared to an OECD average of 36 per cent. Perhaps more significantly, the higher price of oil, coupled with technological advances developed initially in the Gulf of Mexico, allowed the development of the oil and gas reserves of the UK's North Sea continental shelf. Thus, in a short period of time the UK became a net exporter of oil and then gas. With already substantial coal-based electricity and nuclear generating capacity, in the 1980s the UK benefited from an abundance of energy, prices were low; over-capacity was a problem, but security of supply was not an issue and only limited new investment was required to meet growing demand. The government of the time focused on privatisation and liberalisation, which required the creation of a complex system of regulators and watchdogs to ensure that competition promoted efficiency and delivered affordable energy to the UK consumer. As we shall see below, these developments resulted in a significant change in the UK's energy mix, as the coal industry declined, to be replaced by natural gas from the North Sea. This transformation was not just due to the availability of a new energy source: according to Helm (2003: 67), Margaret Thatcher saw the National Union of Mineworkers (NUM) and its leader Arthur Scargill as 'the embodiment of a socialism she was determined to consign to history'. When 'coal was king' the mineworkers were all powerful, and in the early 1970s a series of strikes had led to the three-day working week and the fall of the Conservative government of Ted Heath. But, when the mineworkers went on strike in 1984–85, the coal industry was already in decline and their eventual defeat spelt the end of domestic coal production as the cornerstone of the UK's energy system, as well as economic hardship for those regions of the UK dependent on the coal industry. As subsidies were removed, coal could not compete with natural gas, which prompted the so-called 'dash for gas'.

Since 1970, the share of coal in the UK's primary fuel consumption has fallen from 47.1 to 16.9 per cent in 2008 (see Table 14.2). In 1970, coal provided two-thirds (67.4 per cent) of the fuel input into UK electricity generation, with oil making up two-thirds of the rest – by 1999 coal's share had fallen to 32 per cent (BERR, 2008a: 376–77). In 1992 the share of gas in electricity generation was only 2 per cent, by 1999 its share exceeded that of both coal and nuclear and reached 34 per cent; in 2008 its share stood at 46.1 per cent (DECC, 2009a: 15) and would have been even greater had not high gas prices resulted in substitution of gas for coal (despite the negative environmental impacts). The relative role of oil has not changed substantially since the mid-1980s, and the role of nuclear power has declined due to the retirement of capacity and the relatively low efficiency of the remaining

Table 14.2 UK inland consumption of primary fuels, 1970–2008 (percentage share, energy supplied basis)

Fuel	1970	1980	1990	2000	2007	2008
Coal	47.1	35.8	31.3	16.5	18.1	16.9
Petroleum	44.0	37.5	36.1	32.5	33.4	33.2
Natural gas	5.4	21.9	24.0	40.9	39.8	41.4
Nuclear electricity	3.3	4.8	7.6	8.4	6.2	5.3
Hydro electricity	0.2	0.2	0.2	0.2	0.4	0.5
Net electricity imports	n/a	n/a	0.5	0.5	0.2	0.2
Renewables & waste	n/a	n/a	0.3	1.0	1.9	2.4

Source: Department of Energy and Climate Change, 2009b: 150–53

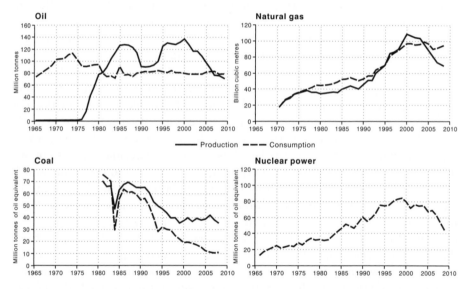

Figure 14.1 Trends in the UK's major energy sources (Produced from data in BP, 2009)

fleet of stations. In 2008 nuclear power only supplied 12.5 per cent of the UK's electricity.

The trends in the production and consumption for the major fuels are shown graphically in Figure 14.1 and they demonstrate the scale of the challenge now facing the UK. The graphs make clear the fact that the 1980s and 1990s were a period of abundance; but oil production peaked in 1999 and gas in 2000, with the UK becoming a net importer of gas in 2004 and oil in 2005. In 2001 the level of coal imports exceeded the level of UK production for the first time, and imports have continued to grow ever since (BERR, 2008b: 22). By 2025 all but one (Sizewell B, which went into service in 1995) of the currently operating nuclear power stations will no longer be in service.

Securing energy for the future

The 2007 Energy White Paper (DTI, 2007: 19) identified two main security of supply challenges:

- increasing reliance on imports of oil and gas in a world where energy demand is rising and energy is becoming more politicised; and

- the requirement for substantial, and timely, private sector investment over the next two decades in gas infrastructure, power stations, and electricity networks.

In the introduction to his report on UK energy security, Malcolm Wicks (2009: 1) points out that 'In recent years Britain was self-sufficient; today we are net importers of over 25 per cent of our annual demand; and by 2020 this proportion will be considerably higher. Estimates of import dependency by 2020 range from 45 per cent or much higher, 70 per cent or more.' In the past the UK government seemed confident that international markets could guarantee the supply of energy. However, the changing geography of oil and gas production; the emergence of new centres of demand, such as China and India, and the rise of 'resource nationalism' and the manipulation of energy exports for geopolitical gain mean that there are substantial risks associated with higher imports of fossil fuels (see Bradshaw, 2009). Helm (2009) maintains that insufficient political capital has been invested in developing and maintaining relations with energy exporting states and in the shaping of the EU's Energy Strategy. Consequently, the UK is ill prepared to face the geopolitical and economic challenges associated with becoming a major energy importing country. For the moment, the UK is in a relatively secure position – the level of imports is modest and easily obtained via international markets. The economic crisis of 2008–9 has dampened energy demand, but, in the medium-term, import dependency is still set to increase at a time when many think global oil production will soon peak and competition for access to natural gas may increase. In the future, the supply of reliable and affordable energy for the UK economy will be no easy matter.

As the 2007 White Paper acknowledges, energy security is not just a matter of matching domestic production and imports to meet demand. In the current context, it also requires substantial investment in new infrastructure to replace ageing generating capacity, facilitate imports, provide storage and enable the transfer of energy to customers. In the past, when the energy system was owned by the state, the government determined what was needed to meet future demand and the state monopolies were funded to develop that infrastructure. Today, the UK government must rely on legislation and regulation to send the right signals to the market, and on private companies to make the necessary investments. While the UK government has spent a great deal of time reviewing and revising its energy strategy, it is questionable whether the market alone can be relied on to deliver the necessary energy revolution.

Helm (2009: 16) maintains that 'The investments have not been made: there is little gas storage; renewables investment has been so low as to merit a lower European-determined target for Britain [15 rather than 20 per cent]; nuclear plants are closing; and the networks are ill placed to tackle the investment agenda.' He goes on to state that 'Britain has been almost completely unprepared for the new energy dependency it is now confronted with' (Helm, 2009: 22). The current economic downturn and the associated 'credit crunch' can only exacerbate the situation as it is now even more difficult to raise capital for very expensive energy projects that have an uncertain return on investment. It is therefore not surprising that there are now calls for greater direct state involvement in the UK energy sector. Lord Browne of Madingley, the former head of BP, in a speech at Cardiff University on 29 March 2009 said:

> 'Competition has been the guiding star of UK energy policy since the 1980s and it worked well while there was a surplus of energy infrastructure capacity. But price competition is now failing to deliver the new, more diversified infrastructure that we urgently need to bolster energy security and meet our climate change targets.'

He went on to say:

> 'I remain convinced that the market is the most effective delivery unit available to society. But the market will need a new strategic direction and a new framework of rules, laid down by government.' (Quoted in Rusbridger and Adam, 2009)

There is another determining factor in meeting the energy security challenge and it is climate change. New investment must not only secure reliable and affordable energy, it must do so in a way that substantially reduces CO_2 emissions; this places major constraints on energy strategy and the means by which energy security can be achieved. It is in this context that renewable energy is seen as a desirable alternative to fossil fuels, as it relies on natural energy flows and sources in the environment which, since they are continuously replenished, will never be exhausted. The UK government has very ambitious plans for the development of renewable energy. *The UK Renewable Energy Strategy* published by DECC (2009c) in July 2009 sets out how the UK will achieve its legally-binding target to ensure that 15 per cent of our energy comes from renewable sources by 2020, which will require a seven-fold increase in the share of renewable energy in little more than a decade!

Transition to a low-carbon UK energy economy

As evidenced by the numerous policy documents and ambitious targets, the UK government is determined to provide global leadership in the introduction of measures to combat climate change. The UK is subject to a Kyoto Protocol target of a 12.5 per cent reduction in 1990 levels of GHG emissions by the period 2008–12. At the same time, the UK is subject to a EU reduction target of 20 per cent of 1990 levels by 2020 (30 per cent if a post-Kyoto agreement is reached). Above and beyond this, prior to coming to power, the Labour government of Tony Blair committed to a national target of a 20 per cent reduction in CO_2 by 2010, which is clearly not going to be met. The 2003 Energy White Paper (DTI, 2003) adopted the recommendation made by the Royal Commission on Environmental Pollution in 2000 and set a longer-term goal to put the UK on a path to reduce CO_2 emissions by 60 per cent by 2050, with real progress by 2020. Most recently, the newly created Committee on Climate Change, an independent body established under the Climate Change Act to advise the UK government on setting carbon budgets and to report to Parliament on the progress made in reducing GHG emissions, advised in its first report that the longer-term target should be a 80 per cent reduction in GHGs by 2050. This is now enshrined in the 2008 Climate Change Act. In the April 2009 Budget, Chancellor of the Exchequer Alastair Darling announced the UK's first Carbon Budget, which set a target of a 34 per cent reduction in CO_2 emissions below the 1990 level by 2020. The second report from the Committee on Climate Change (2009) has made it clear that there has been limited implementation of the measures required to improve carbon efficiency and that a step change is required to meet the emission cuts planned by the government.

Trying to make sense of the plethora of constantly shifting targets is not easy. It is complicated further because some targets are just for CO_2 and some are for all GHGs, a difference which is significant as the energy sector is the major emitter of CO_2. The ways in which emissions are calculated also differs, but the primary focus of the Kyoto Protocols is on the production of GHGs by each state. The UK is committed to very ambitious targets driven by the belief that the dangers of climate change are greater than previously assessed and that it is cheaper and more effective to act sooner rather than later. The 80 per cent target is also seen as an appropriate UK contribution to a global strategy aimed at trying to limit the global temperature rise as close as possible to 2 °C. Clearly, climate change is an extremely complex issue, and one that is beyond the scope of this chapter; the key issue here is that the UK government is committed to dramatic reductions in GHG emissions over a relatively short period of time. This has major implications for the UK's economy in general and the energy sector in particular. The remainder of this section addresses two questions: what is the current state of the UK's GHG emissions, and how will the proposed reductions in emissions impact on the energy sector?

According to UK government data, in 1990 total UK GHG emissions, measured in millions of tonnes of CO_2 equivalent emitted per year (MtCO$_2$e/yr), were 779.9 MtCO$_2$e/yr, resulting in a Kyoto target of an average of below 682.4 for the commitment period of 2008–12 (DECC, 2009e). Trends in GHG emission are shown in Figure 14.2: in 2008 net CO_2 was 531.8 MtCO$_2$e/yr, or

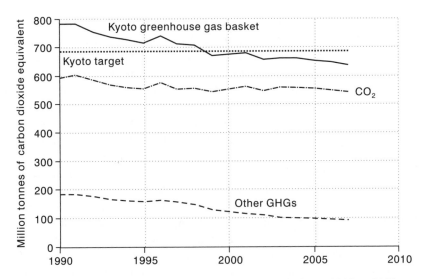

Figure 14.2 UK greenhouse gas and carbon dioxide emissions, 1990 to 2008 (provisional) (Produced from data in DECC, 2009e)

10.3 per cent below 1990 levels (DECC, 2009e). Given that the current recession, and its associated reduction in energy demand, is likely to depress emissions further, the UK seems on track to meet its Kyoto target. How has this been achieved?

On the demand side, the UK has become more energy efficient: by 2000 the energy/GDP ratio had fallen to 60 per cent of the 1990 level and increased energy efficiency is expected to account for 40 per cent of the UK's total carbon savings by 2010 (DTI, 2007: 48). In addition, there are two key 'structural' drivers that account for a large part of the UK's reduction in GHGs since 1990: the change in the energy mix away from coal to natural gas and the de-industrialisation of the economy. The latter has contributed because energy-intensive manufacturing activities have closed, and the goods they produced are now imported; consequently, the economy has become more biased towards service sectors that are far less energy intensive. At the same time, since 1990, CO_2 emissions from coal and other solid fuels have fallen by 46 per cent, though emissions from oil were up 14 per cent and gas 81 per cent (DECC, 2009a: 17). But a modern gas-fired power station, particularly if it is combined heat and power (CHP), produces far less CO_2 per unit of output. During this same period, emissions from transport and from households increased. The figures presented above do not take account of the effect of emissions trading through the EU's Emissions Trading Scheme (ETS), whereby certain installations have to purchase allowances to cover any emissions in excess of their CO_2 allocations, the so-called 'cap-and-trade' system. Data for 2007 show that the UK economy was a 'net acquirer', meaning allowances were purchased to enable emissions above permitted levels. What these figures suggest is that progress so far has been achieved by continuing incremental improvements in energy efficiency and structural changes in the economy and in the energy mix and not by significant behavioural changes on the part of wider society.

In their letter to the Secretary of State for Energy and Climate Change recommending the 80 per cent reduction target by 2050, the Committee on Climate Change suggested a range of options for reducing emissions including improvements in energy efficiency in building and industry and the decarbonisation of the power, transport and heat sectors, and the decarbonisation of industry through the introduction of new technologies such as carbon capture and storage (CCS). Many of these measures figure in the 2007 Energy White Paper, and the Energy Act 2008 and the Climate Change Act 2008 aim to create the legislative and regulatory conditions required for a transition to a low-carbon economy. The government has also published *The UK Low Carbon Transition Plan: National Strategy for Climate and Energy* (DECC, 2009d), which sets out a route-map for the UK's transition to 2020.

The strategy for the future is driven first and foremost by substantial increases in energy efficiency. The government, rightly, sees efficiency and demand reduction as the most cost effective way of addressing the climate change challenge. In the energy sector itself, it seems there are three ways forward: the development of renewable energy; the construction of a new generation of nuclear power stations and the development of CCS technologies to enable the cleaner combustion of fossil fuels and carbon-intensive industrial processes, such as iron and steel and cement production. None of these developments is unproblematic. In the case of renewables, the UK's track record so far is far from convincing. In 2007, renewables accounted for 5 per cent of the electricity generated in the UK, 3.68 per cent from wind, wave, solar and biomass and 1.28 per cent from hydro-electricity (BERR, 2008b: 29). This is up from 1.82 per cent in 1990, but there is a long way to go to meet the EU target of 15 per cent by 2020, and much further to go if, as planned, renewable energy is to make a major contribution to the decarbonisation of the power sector. The potential is greatest in the areas of wind and tidal power, but the change to a more 'distributed' energy system, away from a centralised system focused on a small number of large-scale power producers, poses major technical challenges and considerable cost, as does the intermittent nature of renewable power generation. Power generators are now required to increase their uptake of renewable energy via the Renewables Obligation (RO) that sets targets for the share of renewables in electricity generation. Given this slow progress, it is perhaps not surprising that the UK government has reconsidered the role of nuclear power. In their 2003 Energy White Paper (DTI, 2003), the government put off a decision on nuclear power until a later date. In their 2007 White Paper (DTI, 2007) they concluded that nuclear power now has a place in the transition to a low-carbon economy. A consultation exercise was conducted in 2007 and a White Paper published in January 2008 (BERR, 2008c). In the White Paper (2008c: 5) the government concluded that nuclear power is: low-carbon, affordable, dependable, safe and capable of increasing diversity and reducing our dependence on any one technology or country for our energy supplies. As a result, the government now believes 'that it would be in the public interest to allow energy companies the option of investing in new nuclear power stations; and that the government should take active steps to facilitate this' (2008c: 7). This is a highly contentious decision and there is a long list of conditions attached to the government's willingness to allow the construction of a new generation of nuclear power stations. Not surprisingly, environmentalists do not accept the government's conclusions and maintain that the nuclear option is not necessary to meet the UK's emission targets. At present, there are no concrete proposals to build new nuclear power stations in the UK.

Carbon capture and storage is the last piece of the low-carbon energy puzzle. This is a technology that 'involves the capturing of carbon dioxide emitted when burning fossil fuels, transporting it and storing it in secure spaces such as geological formations, including old oil and gas fields and aquifers (natural underground reservoirs) under the seabed' (BERR, 2006: 172). At present, there is no commercial-scale power station with CCS. In the spring of 2009 the UK government announced that it would back the construction of additional 'clean' coal-fired power stations and that Britain should lead the world in the development of CCS technology. In a global context, the refinement of CCS is seen as critical to help states such as China and India (and the US) where the vast majority of their growing electricity generation is met by burning coal. As part of its energy strategy, the government is supporting the wide-scale deployment of CCS in the UK. Obviously, the decarbonisation of electricity generation is critical to initiatives to promote the use of hybrid and electrical vehicles.

Conclusions

In recent years, the UK government has developed a strategy full of good intentions, but it remains unclear if the necessary investments can be secured and, even if they can, the strategy is based on technologies (with the exception of nuclear power) that are not yet proven at the scale needed to replace the current energy supply system. Added to which, the lead times required to build new power stations and generators, liquefied natural gas (LNG) terminals, pipelines, power grids and storage facilities and the length of their operating lives are such that decisions and commitments need to be made now. Equally, the climate change policy makers maintain that action taken early will be less costly and more effective than adopting a 'wait and see attitude', but the wrong decisions now will lock us into an energy supply system that cannot deliver a low-carbon economy. Thus, the UK's energy sector faces some difficult and profoundly important decisions and the timing could not be worse. In the current economic situation, it is difficult to see how the private sector can commit to the levels of investment required to bring about the necessary energy revolution. As a result, the state may yet have to play a more direct role if the UK is to overcome its energy dilemmas. Whatever the outcome, regardless of what the politicians might say, a low(er) carbon future means higher energy costs in the future and significant changes in the way that we generate and consume energy, and that will have significant impacts on the economy geography of the UK.

Further reading

- DTI (2007) is the definitive analysis by the government of the energy challenges faced by the UK and the strategy needed to address the key issues of energy security and climate change. See the website of DECC for subsequent developments in government policy: www.decc.gov.uk.
- Helm (2003) provides a critical assessment of the changing role of the State in the UK's energy economy from the Conservative Government of Margaret Thatcher through to the Labour government of Tony Blair.
- Wicks (2009) provides a thorough assessment of trends in the global energy economy and the challenges facing the UK, but still places great faith in the ability of the market to provide secure and affordable energy. The report can be obtained from the DECC website at: www.decc.gov.uk/en/content/cms/what_we_do/change_energy/int_energy/security/security.aspx.

References

BERR (2006) *Energy Trends, September 2006*, Department for Business Enterprise & Regulatory Reform. London: TSO.

BERR (2008a) *Digest of United Kingdom Energy Statistics 2008*, Department for Business Enterprise & Regulatory Reform. London: TSO.

BERR (2008b) *UK Energy in Brief July 2008*, Department for Business Enterprise & Regulatory Reform. London: TSO.

BERR (2008c) *Meeting the Energy Challenge, A White Paper on Nuclear Power January 2008*, Department for Business Enterprise & Regulatory Reform. London: TSO.

Beynon, H., Sadler, D. and Hudson, R. (1991) *A Tale of Two Industries: The Contraction of Coal and Steel in the Northeast of England*. Milton Keynes: Open University Press.

BP (2009) *BP Statistical Review of World Energy* (Full workbook). London: BP. Available at www.bp.com/statitiscalreview, accessed 2.11.2009.

Bradshaw, M.J. (2009) The geopolitics of global energy security, *Geography Compass*, 3: 1–18.

Bridge, G. (2009) Material worlds, natural resources, resource geography and the material economy, *Geography Compass*, 3: 1217–44.

Committee on Climate Change (2009) *Meeting Carbon Budgets – The Need for a Step Change*. London: Committee on Climate Change.

Cumbers, A., Mackinnon, D. and Chapman, K. (2003) Innovation, collaboration and learning in regional clusters: a study of SMEs in the Aberdeen oil complex, *Environment and Planning A*, 35: 1689–1706.

DECC (2009a) *Energy Trends, March 2009*, Department of Energy and Climate Change. London: TSO.

DECC (2009b) *Digest of United Kingdom Energy Statistics 2009*. London: TSO, 150–53. Available at: www.decc.gov.uk/en/content/cms/statistics/publications/dukes/dukes.aspx., accessed 2.11.2009.

DECC (2009c) *The UK Renewable Energy Strategy*. London: TSO.

DECC (2009d) *The UK Low Carbon Transition Plan: National Strategy for Climate and Energy*. London: TSO.

DECC (2009e) *UK Climate Change Sustainable Development Indicator: 2008 Greenhouse Gas Emissions, Provisional Figures: Annex A UK Greenhouse Gas Emissions 1990–2008 (provisional) headline results*. Statistical Release, 25 March. Available at www.decc.gov. uk/en/content/cms/statistics/climate_change/climate_change.aspx, accessed 2.11.09.

DTI (2003) *Our Energy Future: Creating a Low Carbon Economy*, White Paper, Department of Trade and Industry. London: TSO.

DTI (2007) *Meeting the Energy Challenge: A White Paper on Energy*, Department of Trade and Industry. London: TSO.

Helm, D. (2003) *Energy, the State, and the Market: British Energy Policy Since 1979*. Oxford: Oxford University Press.

Helm, D. (2009) *Credible Energy Policy: Meeting the Challenges of Security of Supply and Climate Change*. London: Policy Exchange. Available at www.policyexchange.org.uk.

IEA (2007) *Energy Policies of IEA Countries: The United Kingdom 2006 Review*. Paris: IEA.

Rusbridger, A. and Adam, D. (2009) State intervention vital if Britain is to meets its green energy targets, says former BP boss. guardian.co.uk, 25 March. Available at www.guardian.co.uk/environment/2009/mar/25/clean-energy-uk-browne, accessed 8.04.09.

Wicks, M. (2009) *Energy Security: A National Challenge in a Changing World*. London: DECC.

PART 4

LANDSCAPES OF SOCIAL CHANGE

15

RESTRUCTURING UK LABOUR MARKETS: WORK AND EMPLOYMENT IN THE TWENTY-FIRST CENTURY

Kevin Ward

AIMS

- To introduce the notion of labour as a unique form of commodity, and to stress the importance of geography in the constitution of labour markets

- To outline the main contours of the UK labour market's transformation over the last three decades

- To use the growth of the temporary staffing industry as an example of how the UK labour market has become more 'flexible'

My father left school at 15. This was not uncommon for young working-class men. In the early 1960s there were a number of employment options for this social group, particularly in the booming and 'swinging' south-east of England. He chose plumbing. While many of my dad's male friends went into similar manual occupations, most of his female friends, including my mother, left school at the same age and entered relatively low-end, unskilled service jobs. Cashiers, hairdressers and office secretaries were some of the jobs that young working-class women did when they left school with no formal qualifications (McDowell, 2009). Jobs were then clearly gender coded: men worked in manual occupations, women did not; men received a living wage while most women did not; men worked in unionised workplaces with handsomely rewarded overtime remuneration while most women did not. While of course there were exceptions, there were not many. My father's and my mother's routes into the labour market were typical of that particular era.

In other UK regions at the time the choices were perhaps even starker. Men in the north-east or north-west of England who occupied the same class position as my father would have been likely to gain employment in either the coal, manufacturing or steel sectors. For women in these regions their options would have been less affected by the geography of the UK economy; low-end service sector was coded as women's work across most of the UK in the period.

Not surprisingly, much has changed in the UK labour market and its geographical organisation over the last five decades, although the South-east where both my parents first worked remains the economic centre of the country. Many labour-intensive industries, such as coal and steel, however, have ceased to be big employers, meaning that the older industrial regions have suffered high levels of unemployment since the early 1980s (Hudson, 2005). In their place, employment has grown in service sectors such as banking, finance and retailing. The geography to this post-industrial economic growth has been quite different to the geography of deindustrialisation. Cities such as Bristol, Leeds and Manchester, together with London, have seen the numbers of their workers employed in producer services expand over the last two decades (as shown in Chapter 11).

The labour markets my father and mother entered in the mid-1960s, then, were quite different to those of today; from the ways in which people found jobs to the role of the nation-state in the regulation of employment conditions, from the extent to which workers were represented through a union to the social composition of the workforce, from the kinds of jobs people did through to the relationship between employers and employees. These changes and their various geographical dimensions form the focus of this chapter. The first section outlines how we might understand labour as a particular type of commodity to be bought and sold. The second section provides an overview of some of the main trends in the UK labour market over the last few decades. The chapter argues that what has been produced is a labour market that is *both more flexible and more insecure*. The rise of the temporary staffing industry in the UK since the late 1980s is emblematic of these two tendencies, and is discussed in the third section. The final section of this chapter turns to think about some of the future challenges facing the UK labour market and its workers.

Placing labour

An insightful way of thinking about labour is as a *commodity*. As a form of a commodity, labour has a 'use-value' (that is, a practical function, which in my father's case as a young apprentice was taking out old heating systems) and it has an 'exchange-value' (that is, a price, which in my father's case in 1963 was £4 a week). The use-value of workers is their capacity to work – to undertake certain activities and tasks at certain skill levels for certain periods of time – in return for which they are remunerated by employers, most commonly in the form of a weekly wage or a monthly salary. Of course waged workers are unlike any other commodities, however. Compare a worker with the rubber that goes into making the washer on a bathroom tap: the latter is a 'real' commodity – labour on the other hand is not. Rather, people are 'sentient, thinking human beings, conscious agents with their own agendas, pathways and plans' (Hudson, 2005: 3).

Conceiving of labour in this way means that workers are only *temporarily* commodities. Each working day they assume the *form* of a commodity, but this does not change that they were not born and raised to *be* commodities. Waged-workers, because they are people and not bathtubs or toilets, require all manner of things in their lives: entertainment, food, happiness, heat, shelter and so on. There is a whole social and spatial infrastructure that underpins the presence of workers in the labour market. The same cannot be said for shower hoses! And because waged workers are physiologically and psychologically complex beings capable of independent thought and action, they have *agency*. They are not passive objects. Workers can act on their own or with others to change their lives. They necessarily enter into a *social relationship* with their employers which, using the language of Marx, is a *class* relationship. Workers' pseudo-commodity status distinguishes them as a social group from the relative minority of capitalists who purchase their labour power.

This relationship exists inside and outside of the place of work, in the productive and the socially reproductive sphere, and in combination defines the *labour market*. This is not a spot market, however, as neo-classical economists would have us believe, with the price reflecting the supply of workers on the one hand, and the demand for workers on the other. Rather, the labour market is something 'continuously constructed and reconstructed through the very processes that take place within and between them' (Martin and Morrison, 2003: 8). The mediation of supply and demand is then subject to constant struggle (Lier, 2007). Moreover, labour markets are by their very nature 'local'. Making this claim is more than just about arguing that labour markets vary over space, which of course they do – there is more to the argument. It is also important to understand the geographical variability in the processes that are behind the constitution of the 'local' labour market (Peck, 1996). This is about understanding space as both constitutive and generative in the production, regulation and sustaining of labour markets. To use that old geographical cliché, space matters in the theorisation of labour markets.

Restructuring the UK labour market

My father has been self-employed for most of his working life – only in the last five years has he been put on an employment contract. This sets him apart from many of the male workers of his generation. Although they may not have known it at the time, many of those working in the UK labour market up until the 1980s had it rather good. Large national firms had internal labour markets. Think of the likes of British Gas, British Rail, British Telecom, as well as the small number of large banks at the time such as Barclays and Lloyds. This was an era before national and urban economies were punctured by transnational corporations and when foreign ownership was very much in the minority. This meant that for a considerable number of men their first job might also be their last. It was a 'cradle to the grave' type of employment relationship. Of course not everyone experienced the UK labour market in this way. Some workers would move around, from one firm to another. Others, like my father, would work in an occupation in which job tenure was far more precarious. And the women who were in paid employment faired less well. They did not benefit from employment conditions that consisted of permanent employment contracts, a steady career progression from low-skilled to high-skilled posts, transparent and integrated pay structures and regular internal training, even though they may have worked for firms that offered these.

Returning to the many male workers that made up the bulk of the workforce during this era, the expectation was that they had a 'job for life'. Once inside the firm steady progression was possible, and promotion would occur slowly but surely. Now, though, firms with regulated job and pay ladders for each group of workers, internal training provision and clear rules regarding job security are few and far between (Beynon *et al.*, 2002). Since the early 1980s there has been a series of changes in the UK labour market, some stemming from within the UK, others with their origins elsewhere. Some are limited just to the UK, while many others are examples of much broader shifts in the organisation of work across many already industrialised nations. For example, decisions made from elsewhere that have mattered to the UK labour market included those stemming from Brussels (the European Union headquarters) in terms of changes in labour law, or from Tokyo (Nissan headquarters) in terms of the decision to make batteries for its electric cars in its Sunderland plant. If it was ever appropriate to understand the UK labour market as a closed and bounded entity, then that time has surely passed. Rather, the last four decades appear to have revealed how territories such as 'national' labour markets are in fact combinations of 'urban', 'regional', 'national' and 'international' elements coming together to give the impression of being a coherent 'thing'. More specifically, they illustrate how the UK is perhaps most usefully understood as a nationally-distinctive and internally variegated labour system.

These changes have had the consequence of unsettling many of the norms that emerged in the post-Second World War period in the UK and in labour markets in much of the industrialised world. Atkinson (1984) developed the notion of the 'flexible firm' to capture some of these aspects, as set out in Figure 15.1. This model reflects the changes that have been taking place in the nature and composition of the UK workforce, introducing the notion of 'core' and 'peripheral' workers. Core workers are understood as a permanent component of a firm's workforce. They deliver functional flexibility through their capacity to undertake a wide range of tasks. These workers are those

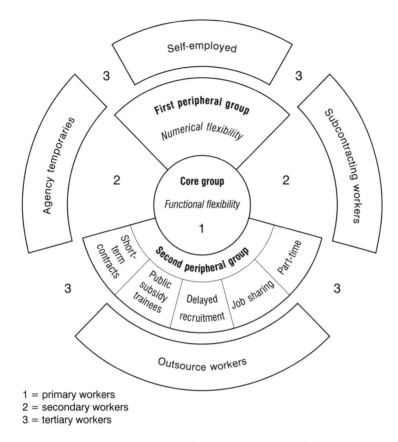

Figure 15.1 The 'flexible firm' (Adapted from Peck, 1996: Figure 3.1)

who have relatively 'sheltered' employment – like many male workers during the 1960s and 1970s. More likely to be permanently hired full-timers with pensions, and with the possibilities of internal progression over the life course, they occupy the firm's 'primary labour markets'.

In contrast in this model, peripheral workers provide a firm with numerical flexibility. Their numbers are increased or reduced as the firm responds to the prevailing labour market conditions. In the current economic crisis, for example, we have seen the ending of placements for temporary agency workers as firms make use of this numerical flexibility (Coe et al., 2009). The secondary labour markets these workers inhabit are ones in which the worker's experiences and the firm's expectations are quite different. In peripheral labour markets workers have a more precarious (and insecure) attachment to the firm and the relationship between employer and employee is a less stable one. In this group are included a range of different types of 'non-standard' employment groups, including part timers, self-employed workers and temporary agency workers. While these sets of workers might not be without their differences in terms of educational attainment, hours worked and so on it is their similarly loose attachment to the core labour market that this model emphasises.

In the remainder of this section I highlight three tendencies behind the emergence of the flexible firm and the flexible labour market. The first is the decline in the number of workers in the UK labour market being recruited directly into permanent posts. In the past few decades, a stable employment relationship between firm and worker was a necessary partner to conditions of skill specificity and on-the-job training. For the worker, permanency of employment provided a basis for accepting relatively low wages in the initial period of employment in return for reasonably good prospects of career and pay promotion in the future. For the firm, hiring workers onto permanent rather than temporary contracts was regarded as an efficient way of ensuring that they benefited from the increased productivity of the fully trained employee. So, as the worker got better at the job – whether it was entering data on a spreadsheet or producing the accounts for a client – the firm was able to gain the benefits. What has emerged in the UK since the 1980s, however, has been a decline in the numbers of workers hired on permanent contracts, while other forms of employment have grown in numbers. It is argued that firms have recruited 'periphery' workers, such as those through temporary staffing agencies, in order to allow them to adapt more quickly to 'external' changes in the demand for products and services, while restricting traditional practices of in-house training and steady promotion to a limited 'core' workforce. 'Flexibility' has become the new buzzword. This had the consequences of producing a highly segmented workforce. Table 15.1 details the changing composition of the UK workforce by employment type over the period 1988–2009; three features are worth highlighting. First, there has been a significant growth in the numbers of part-timers. In the past this has been seen as women's work (O'Reilly and Fagan, 1998), and it still is for the most part. However, the number of men in part-time employment doubled over the period. Second, although 'temporary employment' as captured by official national statistics reveals unspectacular growth over the three decades – hovering around 5–6 per cent of men and 6–7 per cent of women – other evidence suggests an increasing use of this form of labour market intermediary (Ward, 2003). Third, recent years have seen a slight increase in self-employment. Some of this has been 'voluntary', i.e. workers choosing this form of employment for a variety of work and non-work related reasons. In other cases, however, workers have been compelled to register as self-employed so that firms can displace onto the worker the costs of healthcare, insurance and pensions.

The second change in the UK labour market has been in the traditional career structures. Within the classic internal labour market, job ladders were highly regulated for particular groups of jobs, typically establishing different patterns of internal mobility for different occupational groups. A vertical line of career progression was established where work within one job provided the required experience and level of skill for the job at the next level. This structure generated 'natural skill progressions' among jobs in a cluster and ensured that workers were fully trained by the time they were promoted. For

Table 15.1 Full-time, part-time, self- and temporary employment in the UK labour market, 1988–2009 (thousands)

Year (April–June)	All people in employment	Male employment	Male full-time employment	Male part-time employment	Male self-employment	Male temporary employment	Women in employment	Women full-time employment	Women part-time employment	Women self-employment	Women temporary employment
1988	25,970	14,790	N/A	N/A	N/A	N/A	11,180	N/A	N/A	N/A	N/A
1990	26,935	15,192	N/A	N/A	N/A	N/A	11,742	N/A	N/A	N/A	N/A
1992	25,601	14,119	13,129	990	2,567	543	11,481	6,452	5,030	907	731
1994	25,448	13,886	12,814	1,071	2,577	657	11,562	6,402	5,160	906	832
1996	26,009	14,163	12,946	1,217	2,584	739	11,845	6,548	5,298	913	912
1998	26,715	14,571	13,272	1,299	2,450	768	12,145	6,718	5,427	913	946
2000	27,469	14,900	13,529	1,371	2,354	777	12,568	6,997	5,572	896	930
2002	27,911	15,073	13,598	1,475	2,437	743	12,838	7,188	5,650	896	847
2004	28,428	15,371	13,745	1,626	2,690	701	13,057	7,324	5,733	980	809
2006	28,998	15,628	13,955	1,673	2,690	661	13,371	7,692	5,678	1,018	805
2008	29,505	15,938	14,132	1,805	2,788	642	13,568	7,881	5,687	1,051	765
2009	28,933	15,489	13,596	1,893	2,733	693	13,444	7,764	5,680	1,085	757

Source: Labour Force Survey, available at www.statistics.gov.uk

the firm, this system was an effective means of capturing the benefits of on-the-job training, as acquired skills are applied to higher-level tasks. For workers, regulated and transparent paths of internal mobility not only facilitated steady career advancement based on the acquisition of skills and experience, but also introduced a degree of certainty about long-term plans and underpinned realistic expectations around improvements in living standards. Arranging mortgages and pensions, for example, was made easier as a result of the relative certainty in predicting future job and pay changes.

Since the 1990s these traditional career structures have been transformed for many workers. Job ladders have been 'de-layered' in an apparent effort to improve communications among workers, and the emphasis of training policy has switched from investment in technical skills to 'generic competencies', reducing the links between vertical levels of jobs. This has meant that workers can no longer expect the training and experience required for one job to have occurred in the job below it. This trend has been particularly pronounced amongst white-collar workers. In some industries, most noticeably banking and finance, the late 1980s and early 1990s saw workforce numbers tumble. Geographically, employment in the centres of cities and towns has been lost. Branches have been closed and the functions transferred either to call centres in the north-east or north-west of England or offshored to India (Leyson et al., 2008).

Third, and finally, recent decades have witnessed an ending of any expectation of a 'job for life' in the UK labour market. While in the past this sense was institutionalised within firms through formal and informal arrangements, as we have seen earlier in this chapter, expectations of job security were turned on their head during the late 1980s and early 1990s. For Elliot and Atkinson (1998) this constituted the onset of 'the age of insecurity'. Responding to pressures to 'downsize' to lower costs, improve productivity and increase profits, firms set about reducing the size of their 'on the books' workforce. In some cases, firms would do this while also rehiring staff on different types of employment contracts, most commonly through temporary staffing agencies. 'Downsizing' worker numbers was understood to reflect a firm's short-run performance. So to appear 'productive' and to meet the aspirations of shareholders and those in the City of London who value corporate shares, many firms set about reducing their workforces. In many cases, as the numbers of workers fell so the share prices rose. British Telecom, for example, 'downsized' tens of thousands of workers in the 1990s, and its stock market value soared! What has been most revealing about the undermining of any sense of a 'job for life' is not that insecurity was introduced into the UK labour market. That has always been there, even during the 1960s and 1970s: in many working-class, blue-collar occupations, such as plumbing, there was no 'job for life' – there was a trade or a career for life, possibly. What sets these changes apart is the extent of insecurity in white-collar occupations, at the same time as existing norms of insecurity down the job ladder were deepened. In the place of a 'job for life' has emerged the language of the so-called 'free agent' worker (Pink, 2001) and the 'boundaryless career' (Arthur and

Rosseau, 1996). The risk of securing and remaining in employment has been increasingly displaced onto individuals – hence the rise in self-employment – with of course some better situated than others to prosper under these new realities. In the next section the chapter turns to the emergence of the UK temporary staffing industry. This has been both a benefactor from, and a causal factor behind, these changes in the UK labour market. As Peck and Theodore (2007: 183) put it, the '[temporary staffing industry's] expansion should be understood both in terms of the changing ways in which client businesses have deployed a growing workforce of agency-supplied temps and the wider managerial imperatives to pare labour costs, redesign job functions and flexibilise employment relations.'

The UK temporary staffing industry: labour market flexibility personified?

The temporary staffing industry has prospered as the UK labour market has been restructured. In turn, the restructuring of the UK labour market has been facilitated by the growth of the temporary staffing industry. Temporary staffing agencies are a form of labour market intermediary, meeting the needs of client firms for contract workers of many kinds. 'With a core business of *labour supply*, temporary staffing is a very particular kind of "people-based" business service activity, and one which, by its very nature, is always *delivered* locally' (Coe et al., 2007: 504). Agencies allow firms a degree of numerical flexibility, so that if economic circumstances change workers contracts can be ended at very short notice. Of course, temporary agency work is not new. For example, *The Economist* (1962: 705–6) reported that at the beginning of the 1960s London had over 320 temporary staffing agencies. Since the entrance in the early 1960s into the UK labour market of the large US temporary staffing agencies – such as Kelly Services and Manpower – temporary agency workers have been used as fill-ins for workers who were ill, on maternity leave or on holiday. At first, however, temporary agency workers were marginal to the main thrust of a client firm's business.

This all began to change in the early 1990s. As Nollen (1996: 567) describes, 'temporaries, although still peripheral workers, are [now] integral to business strategy. Temporary employment is a permanent feature of the business landscape.' Like the clients they supply with workers, the temporary staffing agencies are in it for the money. Ofstead (1999: 292) puts it thus: temporary staffing agencies 'like the businesses they serve ... want to grow and become as profitable as they can'. During the 1990s, temporary staffing agencies in the UK began to diversify their activities. They began to place workers across a wider range of economic sectors, increasing the amount of business they did outside of clerical and light industrial activities, and seeking to move into higher-end occupations and more professional sectors. They also began to offer extra services in addition to the placement of workers such as managing

Table 15.2 Growth in urban England of temporary staffing agencies, 1971–2001

	1971	1981	1991	2001	Per cent change, 1971–2001
Birmingham	60	69	138	293	388
Bristol	26	49	108	212	715
Leeds	35	49	111	225	542
Liverpool	29	32	42	54	86
Manchester	65	89	168	252	287
Newcastle	18	31	78	75	316
Total	233	319	645	1,111	385

Source: Ward, 2005: Table 2

payrolls, screening workers for permanent placements, staffing new corporate start-ups. In this way agencies sought to 'deepen' their relationship with client firms (Ward, 2003).

These changing practices are reflected in the growth in the number of agencies in the UK from a low base at the beginning of the 1970s. In addition to London, many agencies are concentrated in and around some of the UK's largest cities, which have experienced quite spectacular growth in the number of temporary staffing agencies (Table 15.2). The number of workers placed into a job in the UK through a temporary staffing agency is currently 1.2 million per week. And, as the amount of business done by agencies has grown, so has the value of the UK industry – current figures value it at just over £23.1 billion a year. Despite this growth there has been only a gradual change in the perceptions of being 'just a temp' (Henson, 1996). For many workers, gaining employment through a temporary staffing agency remains a constrained choice. Significant numbers would like to be directly and permanently employed by a firm (TUC, 2008). Where for some the flexibility offered through temping is a positive thing, others are willing to trade the flexibility for greater security. Being a temp remains then something to avoid. Despite the campaigns of business groups and trade associations such as the Recruitment and Employment Confederation (REC), it remains something to which a social stigma is attached. Making specific reference to call centres, Figure 15.2 reflects widely held views about the status of the temp in the workplace. Two recent examples reflect the relative importance of temporary staffing agencies in the UK labour market. The first example is the role they played in placing into work A8 migrants who entered the UK post-accession in April 2004 (see Chapter 17). These countries – the Czech Republic, Estonia, Hungary, Latvia, Lithuania, Poland, Slovakia and Slovenia – were those who joined the EU on 1 April 2004 and whose citizens were then allowed to enter and work in the UK. Evidence suggests that agencies were initially important in facilitating the movement of workers into the country and in placing them into employment (McDowell et al., 2008; TUC, 2007). The second example is how in the current economic recession agencies have acted as 'buffers' for what remains of the core workforces at client firms. Last in, first out: the evidence suggests that after years of regular placements, many temporary

Figure 15.2 Representations of 'just a temp' (Courtesy of CallCenterComics.com) Accessed September 2009.

agency workers are currently experiencing the downside of this particular form of labour market flexibility. For example, in February 2009, 850 temporary agency workers lost their jobs at BMW's Cowley plant near Oxford and received no company redundancy pay, highlighting the vulnerability of those not employed as permanent staff. Of the 10,000 workers British Telecom (BT) announced in November 2008 that it was letting go, more than two-thirds were placed into their jobs though a temporary staffing agency. In both the cases of BMW and BT, much of their UK workforce was hired through the US-based temporary staffing agency, Manpower (Bowcott and Hencke, 2009).

Conclusion: what sort of working futures?

This chapter has provided an overview of the changes in the UK labour market. My father – who now works for a large transnational – is about to retire. My brother-in-law works for the same firm as my father, and the plumbing industry he joined in the early 2000s was one far removed from the one my father entered all those years ago. There are now very few apprenticeships, next to no union coverage, much greater self-employment, and many more accreditation systems (such as those offered by CORGI), which require plumbers to update their training on a regular basis. And yet some things have remained relatively unchanged. The hours are still long, the work is still physically demanding, the majority of the workers come from working–class backgrounds and most plumbers are still white and male, despite the changing gender and ethnic composition of the UK workforce. And, of course, this work still takes place within a geographically uneven capitalist system.

It is clear that the UK labour market has undergone tremendous changes over recent decades. Opinions remain divided on their extent and what they mean for different types of places and different sets of workers. North–South divisions still persist. Within 'northern' regions there are areas that have done well and those that have not. Likewise, within 'southern' regions there are areas that have prospered. At best we can say that the UK's labour market geographies remain complex: it is a nationally distinctive and internally variegated national labour system. In terms of the people in the UK labour market, there are now more women in the workforce than ever before. Yet the gender pay gap continues to get worse. Current data reveal that women working full time earn '87 percent of the male median full time hourly wage or just under 83 per cent of the male mean full time hourly wage' (Women and Work Commission, 2006: 1). Women continue to be under-represented in senior management. Some ethnic minorities have done better than others during the last four decades. Black African UK-born males appear to have faired worse, in terms of rates of employment and rates of pay (Dustmann et al., 2003).

If anything has become clear in the last 24 months, it is that capitalism continues to be an incredibly adaptable and robust economic and social system – and a profoundly uneven geographical one to boot. It remains unclear what the longer-term consequences of the current economic juncture will mean for the millions of workers in the UK labour market. According to the UK government, and others such as the Confederation of British Industry (CBI), it is the changes that have been outlined in this chapter, together with others, that are behind the reasonable performance of the UK economy since the beginning of the recession at the end of 2007. The UK's 'flexible' labour market has behaved as one would expect: it has contracted, with those with the loosest attachment to the labour market – such as workers placed through temporary staffing agencies – being the first to be without a job. Those working in UK regions in which are sited more disposable elements of the production network have also found themselves under threat, as transnationals have looked to move plants to locations where the labour costs are cheaper. This reflects a more general, global trend, as employment is offshored. For some, such as the Trade Union Congress (TUC), the same set of changes in the organisation and regulation of the UK labour market have left workers of all sorts horribly exposed. For others, the recent changes represent an opportunity to re-think the relationship between work, economy and the environment in the name of a more just future. In what ways should society reassess how it values different types of work (particularly care work)?

What should the role of waged work be in moving the UK economy to a more carbon-responsible position? What might be done about ensuring all waged workers in the UK earn something approaching a 'living wage'? How might we think ethically and responsibly about working in a globally inter-connected and inter-dependent world? These are just some of the questions that many will be grappling with long into the twenty-first century.

Further reading

- Castree et al. (2004) is an upper-level textbook that makes the case for a labour-centred analysis of contemporary global capitalism.
- Martin and Morrison (2003) is an edited collection of contributions from across the spectrum of human geography, revealing the different way the discipline seeks to understand the concept of 'labour'.
- Peck (1996) is a research monograph that draws on a range of case studies to make the case for the constitutive characteristics of space, place and scale in the formation and regulation of labour markets.

References

Arthur, M.B. and Rosseau, D.S. (1996) *The Boundaryless Career: A New Employment Principle for a New Organisational Era*. Oxford: Oxford University Press.

Atkinson, J. (1984) Manpower strategies for flexible organisations, *Personnel Management*, August, 28–31.

Beynon, H., Grimshaw, D., Rubery, J. and Ward, K. (2002) *Managing Employment Change: The New Realities of Work*. Oxford: Oxford University Press.

Bowcott, O. and Hencke, D. (2009) Fear for deal on agency workers' payoffs, *The Guardian*, 21 February. Available at www.guardian.co.uk/politics/2009/feb/21/agency-workers-payoffs, accessed 7.9.09.

Castree, N., Coe, N.M., Ward, K. and Samers, M. (2004) *Space of Work: Global Capitalism and Geographies of Labour*. London: Sage.

Coe, N.M., Johns, J. and Ward, K. (2007) Mapping the globalization of the temporary staffing industry, *Professional Geographer*, 59: 503–20.

Coe, N.M., Johns, J. and Ward, K. (2009) *The Economic Crisis and Private Employment Agencies: Opportunities and Challenges*. Geneva: ILO.

Dustmann, C., Fabbri, F., Preston, I. and Wadsworth, J. (2003) *Labour Market Performance of Immigrants in the UK Labour Market*. London: Home Office.

Economist, The (1962) Girls for hire, 19 May, pp. 705–6.

Elliott, L. and Atkinson, R. (1998) *The Age of Insecurity*. London: Verso.

Henson, K.D. (1996) *Just a Temp*. Philadelphia: Temple University Press.

Hudson, R. (2005) *Economies Geographies*. London: Sage.

Leyson, A., French, S. and Signoretta, P. (2008) Financial exclusion and the geography of bank and building society closure in Britain, *Transactions of the Institute of British Geographers NS*, 33: 447–65.

224   LANDSCAPES OF SOCIAL CHANGE

Lier, D.C. (2007) Places of work, scales of organising: A review of labour geography, *Geography Compass*, 1: 814 – 33.

Martin, R. and Morrison, P. (eds) (2003) *Geographies of Labour Market Inequality*. London: Routledge.

McDowell, L. (2009) *Working Bodies: Interactive Service Employment and Workplace Identities*. Oxford: Wiley Blackwell.

McDowell, L., Batnitzky, A. and Dyer, S. (2008) Internationalisation and the spaces of temporary labour: the global assembly of a local workforce, *British Journal of Industrial Relations*, 46: 750–70.

Nollen, S.D. (1996) Negative aspects of temporary employment, *Journal of Labor Research*, 17: 567 – 82.

Ofstead, C.M. (1999) Temporary help firms as entrepreneurial actors, *Sociological Forum*, 14: 273 – 94.

O'Reilly, J. and Fagan, C. (eds) (1998) *Part-Time Prospects*. London: Routledge.

Peck, J. (1996) *Work-place: The Social Regulation of Labor Markets*. London: Guilford Press.

Peck, J. and Theodore, N. (2007) Flexible recession: the temporary staffing industry and mediated work in the United States, *Cambridge Journal of Economics*, 31: 171 – 92.

Pink, H. (2001) *Free Agency Nation: How America's New Independent Workers are Transforming the Way We Live.* New York: Warner.

TUC (2007) *Migrant Agency Workers in the UK*. London: TUC.

TUC (2008) *Final report of the Commission on Vulnerable Employees*. London: TUC.

Ward, K. (2003) UK temporary staffing industry: industry structure and evolutionary dynamics, *Environment and Planning A*, 36: 2119–339.

Ward, K. (2005) Making Manchester 'flexible': competition and change in the temporary staffing industry, *Geoforum*, 36: 223–40.

Women and Work Commission (2006) *Shaping a Better Future*. London: Department of Trade and Industry.

16

NEW MIGRANT DIVISIONS OF LABOUR

Jane Wills, Cathy McIlwaine, Kavita Datta, Jon May, Joanna Herbert and Yara Evans

AIMS

- To introduce and describe the new Migrant Division of Labour (MDL) within the UK economy

- To profile the nature of recent immigration reform in the UK

- To explore the interconnections between immigration and the labour market, particularly in London

- To identify important debates concerning the wider implications of the MDL

On 1 January 2008, the bulk of the cleaning services at Queen Mary, University of London, were moved back in-house. Following a successful living wage campaign, university managers were convinced of the need to improve the pay and conditions of cleaners, and they decided to stop using a subcontractor and bring the service in house. Cleaners who used to be employed by an outside company became full members of the University team and they secured significant increases in pay and benefits, as well as having the opportunity for further training and career development. This decision flies in the face of the shift towards subcontracting and agency labour that has been dominant in the UK labour market since the 1980s (see Chapter 15). The decision also provided an opportunity to find out more about the workers doing these jobs.

Interviews with 73 of the cleaning staff during late 2008 revealed that they were born in as many as 24 different countries (see Table 16.1). While a large number were the sole representative of their country of birth, a significant number were found to come from Somalia, with other larger groups hailing from Ghana, Nigeria and the Caribbean. These cleaners were found to bring extraordinary geographical and ethnic diversity to the community

Table 16.1   The country of birth of surveyed cleaners at Queen Mary, University of London

Country of birth	Number	Per cent of those surveyed
Somalia	17	24
Ghana	13	18
Nigeria	9	13
Jamaica	5	7
UK	4	6
Bolivia	3	4
Montserrat	2	3
Morocco	2	3
Philippines	2	3
Bangladesh	1	1.4
Barbados	1	1.4
Colombia	1	1.4
Congo	1	1.4
Dominica	1	1.4
Ecuador	1	1.4
France	1	1.4
Grenada	1	1.4
Ivory Coast	1	1.4
Madeira, Portugal	1	1.4
Romania	1	1.4
St Lucia	1	1.4
St Vincent	1	1.4
The Netherlands	1	1.4
(2 missing data)	71	–

Source: Wills with Kakpo and Begum, 2009: 9

at Queen Mary. Whereas the majority of staff at the University described their ethnicity as white (65 per cent), only 20 per cent of cleaners described themselves in this way – and many of these were from various European countries rather than the UK itself. When asked about their ethnicity, most cleaners described themselves as black, with more than half being black African (39 or 56 per cent of the sample, reflecting those born in Africa as well as a number of European-born cleaners of African parentage) and a further one-fifth being black Caribbean (12 cleaners or 17 per cent of the sample).

Such data provides a dramatic illustration of what we would call the migrant division of labour (MDL) (see also May et al., 2007; Wills et al., 2009, 2010). During the past 20 years or so, the UK has become a country of net immigration, and these new arrivals have transformed the labour force – particularly in a global city like London. While there is a long history of immigration to the UK, recent arrivals have hailed from a greater diversity of countries with all that that implies for cultural and ethnic differentiation. Migrants have also arrived with divergent levels of skill and have used a wider number of immigration channels to reach the UK. The anthropologist

Steven Vertovec (2007) has referred to this new immigration as being character-ised by 'super diversity' and a number of researchers have documented this diversity in low-paid workplaces across the UK (see MacKenzie and Forde, 2009; Mathews and Ruhs, 2007). These processes have contributed to the creation of a MDL in which foreign-born workers are particularly positioned in relation to the low-paid labour market, and particularly so in urban loca-tions. As we outline more fully below, the 'dual framework' whereby wages here are much higher than wages at 'home', coupled with immigration status and its associated limits on access to the benefit system, make migrants more able and willing to take up low-paid employment. This chapter outlines recent changes in the UK's immigration policy and practice before then exploring the implications these changes have for the labour market and the labour supply.

Immigration policy within the UK

Since its election to government in 1997, Britain's New Labour party has reconfigured immigration policy. In contrast to the post-war consensus that sought to limit immigration into the country almost as soon as it started to grow, ministers now extol the economic virtues of immigration and the importance of attracting talent to the UK. The government has sought to open its borders to those who are seen as sufficiently highly skilled and entrepre-neurial to contribute to the wealth of the nation. Rather than limiting immi-gration per se, the government has sought to manage it for economic advantage. As a result, and in contrast to the previous century, the UK has been a country of net immigration since the early 1990s (see Figure 16.1). As indicated, immigration had already started to rise, due to an increase in asy-lum applications and growing numbers of international students during the early years of the decade, but the rate has since accelerated over the years with the shift in government thinking.

The government's new approach to manage migration has culminated in the implementation of a points-based immigration regime that is being rolled out between 2007 and 2010. As outlined in Table 16.2, the highly skilled migrants at the top of this hierarchy (Tier 1) have full rights to the labour market and the benefit system. They are welcome to stay and work for as long as they like, backed up with a pathway to citizenship should they want to remain. Those who are granted access to work for a particular employer in an identified shortage sector (that depends on research and analysis con-ducted by the Migration Advisory Committee) have to have a requisite level of English language skills to fulfil the terms of Tier 2. These workers have no rights to the benefit system, although they do have the right to apply for citi-zenship if they sustain their employment for more than five years. In Tier 3, there is no space for relatively unskilled workers from outside the European

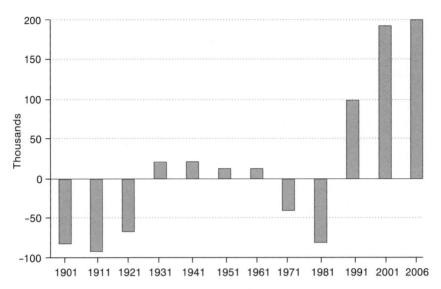

Figure 16.1   Net migration to the UK (ONS, 2007; Wills et al., 2010: 12)
Notes: Includes civilian migration and other adjustments. Ten year averages are used before 1931 and between 1951 and 1970. A 20 year average is used between 1931 and 1950. Data prior to 1971 are for calendar years, data for 1971 onwards are mid-year estimates.

Union (EU) to gain access to employment and citizenship within the UK. Instead, the government now expects all unskilled vacancies to be filled by migrants from the wider EU. Those from outside the EU can only gain access to the UK's labour market through family reunification, as international students or as refugees. Indeed, those running educational courses for international students (under Tier 4) are now expected to sponsor and monitor the activities of their students for violations of immigration control. Students are expected to attend their classes and to limit their working hours to 20 hours a week during term time, and unless they satisfy particular criteria they are required to leave the country once their course is complete.

In many ways, this new system represents little change for the would-be unskilled migrant from beyond the EU. There has been no opportunity for unskilled non-Europeans without family contacts or grounds for asylum to enter the British labour market since the abandonment of the grade C vouchers that were instituted by the 1962 Commonwealth Immigrant Act and changes in citizenship entitlement that were implemented in the late 1960s. The limited sector-based schemes that briefly existed for those working in agriculture, hospitality and food processing under the terms of the Nationality, Immigration and Asylum Act (2002) were abolished in 2005, and in any case these rarely recruited from beyond the wider Europe, including Ukraine. For the past 40 years, so-called unskilled migrants from outside the EU have only been able to enter Britain through family reunification, the asylum system,

Table 16.2 The UK's points-based immigration regime

	Description	Terms of entry
Tier 1	Highly skilled individuals to contribute to growth and productivity	Based on qualifications, previous earnings, age and other criteria. Granted unrestricted access to the labour market and benefits for 2 years with dependants. Extension, settlement and citizenship can follow reassessment.
Tier 2	Skilled workers with a job offer to fill gaps in the UK labour force	Job offer in shortage area or where not displacing a UK/EU worker. Job must be at NVQ3 or above and have been advertised. Employers act as sponsor. The recruit has to meet English language requirements and can only change employer if they reapply for a new permit. If they meet the points demanded by Tier 1, they can subsequently bring dependants and secure unrestricted access to the labour market and benefit system. It is possible to apply for settlement after 5 years in the UK.
Tier 3	Limited numbers of low-skilled workers needed to fill specific temporary labour shortages	Quota-based, operator-led, time-limited schemes will run subject to review involving countries with which the UK has a robust returns arrangement. Expected to include only migrants from the A2 (Bulgaria and Romania).
Tier 4	Students	Dependent upon sponsorship and granted only for the period of the course. Tighter controls over the institutions able to sponsor students. Can work 20 hours a week and full time in holidays.
Tier 5	Youth mobility and temporary workers: people allowed to work in the UK for a limited period of time to satisfy primarily non-economic objectives	Includes Commonwealth Working Holiday Makers scheme and au pairs. For 18–30 year olds for up to 24 months. Can work for 12 of the 24 months. No rights for dependants, no right to switch to a different Tier. Sponsorship required from national governments with agreed returns policy and reciprocal relationships with the UK. Limited additional options for temporary workers in the creative and sporting sectors, for voluntary work, religious activities, international exchange and agreements.

Source: Home Office, 2006

as international students, tourists, or clandestinely in the back of a lorry or under a train.

Thus, although the UK is now more open to some foreigners, selling itself as a cosmopolitan country at the heart of economic globalisation, the nation's borders are not open to all. There is no greater access for the unskilled from outside the EU than there was in the past. However, the country has opened its borders to unskilled immigration from within the EU. Following the accession of Cyprus, Malta, Estonia, Croatia, Czech Republic, Hungary, Lithuania, Poland, Slovakia and Slovenia (the latter eight eastern European countries being known as the Accession 8 (or A8) in May 2004, and the accession of Bulgaria and Romania known as the Accession 2 (or A2) in January 2007, unprecedented numbers of Europeans have arrived to work in the UK (see also Chapter 17). In contrast to the majority of other European nations, the UK did not impose restrictions on the arrival of A8 migrants from May 2004, although greater restrictions have been imposed on those from the A2 countries since January 2007. Official statistics indicate that more than half a million migrants have arrived since this time, transforming low-paid labour markets across the UK. Many of these migrants will have since returned home, particularly since the economic crisis of 2008–9, and these migrants have been more likely to stay for shorter periods than earlier waves of New Commonwealth immigrants who came to Britain after the Second World War, following a government initiative to swell Britain's labour force.

In contrast to the approach towards Europe, however, official policy is now increasingly draconian in regard to so-called unskilled migrants from beyond the EU. As the Home Office puts it, British immigration policy has been developed in the interests of the economy with 'borders that are open to those who bring skills, talent, business and creativity that boost our economy, yet closed to those who might cause us harm or seek to enter illegally' (Home Office and Commonwealth Office, 2007: 2). In this vein, the government has implemented tighter border control via the visa regime, newly introduced biometric identity cards and border surveillance as well as stepping up workplace raids, dispersal, detention and deportation. The UK now has as many as nine Immigration Removal Centres and four Short-term Holding Centres that house failed asylum seekers and irregular migrants, many of whom do not have the papers needed for identification or deportation abroad. While such policies have not yet stopped people trying to enter the UK illegally, with current estimates suggesting that at least 1,000 people a year die trying to reach the EU (Legrain, 2007), the new regime is making it much harder for irregular migrants and their would-be employers.

In addition to their reconfiguration of the immigration regime, the New Labour government has also moved to alter the debate about multiculturalism within the UK. In the wake of increased public concern about immigration and the possibility of home-grown terrorism, debate has shifted towards the importance of shared values and community cohesion. No longer is it deemed desirable for migrants to actively maintain their cultural identities from their home countries alone; they are now required to show that they intend to integrate into a dominant 'British way of life'. This is to be achieved through the acquisition of English language skills, citizenship tests and ceremonies, and there is now a duty on state-funded bodies to promote trans-cultural links.

Immigration and the labour market

Established theoretical analyses of the intersections between immigration and the labour market in the UK were largely developed in the context of rising post-war immigration. Labour shortages after the Second World War meant that countries like Britain needed to attract workers, and in tandem with the development of guest worker schemes in many European nations, the British government endorsed limited recruitment from beyond the UK. While 'natives' were able to secure better forms of employment, immigrants arrived to fill the 'bottom end' jobs as demonstrated by the concentration of African-Caribbean workers in the National Health Service and London Transport, and Indian, Pakistani and Bangladeshi workers in manufacturing jobs, particularly in the clothing and textile trades.

This situation prompted scholars to focus on employer demand as a key determinant of immigration. At a time when Marxist ideas were widely adopted in the social sciences, scholars argued that immigration was functional to capitalism, that it was driven by employer demand for cheap and pliable labour, and dependent upon surplus populations or 'reserve armies' in the former colonial world. In one of the most sophisticated and well-known developments of the arguments from this period, Piore (1979) posited that there was an inherent role for immigrant labour in advanced capitalist economies. He argued that labour-dependent employers in tight labour markets with limited profit margins could not simply increase wage levels to attract 'natives' into the work. Raising wages at the bottom would mean demand for wage rises elsewhere, potentially leading to structural inflation: although this is somewhat questionable after the experience of implementing minimum wage legislation in the UK, there is never a strong appetite for increasing wages for these kinds of jobs.

In addition, however, Piore also argued that employers needed to find workers with the motivation to work. Immigrants fitted this bill as they often arrived with poor language skills and low levels of education with few alternative sources of work. Immigrants were recruited in the wake of the 'natives' who were moving out of such jobs. Moreover, immigrants were argued to be then further confined by racism and wider socio-economic disadvantage to remain in 'bottom-end' jobs. Such analyses were later extended by the notion of the segmented labour market in which a combination of personal characteristics and employer discrimination corralled individuals into particular kinds of employment – producing spatially-differentiated gender and ethnic divisions of labour (Coe et al., 2007; Peck, 1996).

Thus in the UK today, migrants are still concentrated in low-paying jobs that others are unwilling to do. However, immigration is also a feature of highly skilled work. Reflecting changes in the immigration regime, highly skilled workers are being recruited into jobs like finance, law, health and research. As illustrated in Figure 16.2, foreign-born workers are over-concentrated at both ends of the occupational hierarchy – predominating in the professions as well as in elementary jobs. This graph also indicates, however, that A8 migrants – those from eastern Europe – are concentrated in the 'bottom-end' jobs, while non-EU migrants are concentrated in professional jobs. Thus new arrivals from eastern Europe are over-represented in personal services, process/plant and machine operations and in elementary occupations, and they are under-represented in higher-end jobs. In contrast, and reflecting the trend towards stratified immigration control on the basis of skills, foreign-born workers

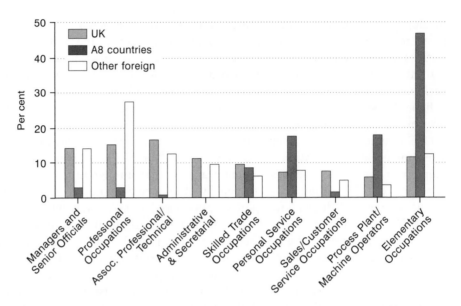

Figure 16.2 The occupational distribution of foreign-born workers in the UK, 2006 (Labour Force Survey, Spring 2006; Wills et al., 2010:14)

from beyond the EU are over-represented in the professions. Yet it is important to note that more fine-grained research points to a different picture at the sub-regional scale. New data collected in London, for example, and outlined further below, shows that more than 50 per cent of the workers in jobs like catering, cleaning and care were foreign-born by the mid-2000s, with significant numbers of workers from Africa, Latin America and the Caribbean (Wills et al., 2009, 2010). Similar trends are likely to be found in other conurbations, albeit on a much smaller scale. In cities where there is already a sizeable diasporic community, and/or universities that attract foreign students, and/or an asylum processing centre, in tandem with employment opportunities, migrant workers are likely to be found in significant numbers today (Philimore and Goodson, 2006).

Migrant workers are attracted into jobs at the 'bottom end' of the employment hierarchy in these locations because the wages appear much better than they do back 'home' and because immigration controls often restrict their access to any alternative state-provided support. In what Waldinger and Lichter (2003) refer to as the 'dual frame of reference', migrants feel better paid than their counterparts back 'home' while also lacking entitlement to the benefits available to citizens. In the context of the welfare state, migrants become particularly valuable to employers as they are often more willing to work in low-paid employment than the entitled 'native' labour supply. In addition, however, some employers may also prefer to employ foreigners as a way to 'distance' themselves from the moral economy of the labour being done. Whereas they might have baulked at employing 'their own kind' to do dirty, dangerous and difficult work for very low rates of pay, it can be easier to employ those who are coded as 'other' (Waldinger and Lichter, 2003: 40).

The literature suggests that employers employ migrants *because* they are migrants, with different qualities to the 'native' labour supply. Moreover, while some employers will have little choice of labour supply, high rates of both immigration and unemployment in many of British cities mean that employers often have some discretion over whom to employ. In the context of over-supply for low-waged labour, employers tend to develop a nationality-based 'shorthand' in their approach to recruitment. In what has been called the 'hiring queue' (Model, 2002), employers tend to adopt national and racialised stereotypes to determine the reliability of potential recruits. In her multivariate data analysis of trends in London, New York and Toronto, for example, Model (2002: 132) found a 'relatively stable cross-national hierarchy of discrimination' which reflected established prejudices and stereotypes as well as the impact of employers' ethnicity and different streams of labour supply. Given the opportunity, employers' preferences in all three cities were for foreign-born whites, followed by Asian immigrants from African countries and then the Chinese. These workers were found to fare at least as well as native-born whites (especially men) in each geographic location. In contrast, however, employers would choose to employ Africans, Pakistanis and Bangladeshis only in the absence of the more highly ranked groups. In the US, African-American men were found to be the least favoured group, selected only when there were no migrants left in the queue.

As employers compete for the most desirable workers, those with the least desirable jobs are likely to have less choice over whom to employ. Hence in the UK, we would expect that the employers with the poorest paying and least desirable jobs would be most dependent upon the 'least desirable' workers who have the fewest alternative sources of income and work. However, the volumes of potential workers looking for work will also play a big part in the changing MDL and employers' hiring queues. Despite the geographic dispersal of many workers from central and eastern Europe, and the government's efforts to disperse supported asylum seekers across the UK, foreign-born workers remain concentrated in cities, and particularly within London. Our analysis of the Labour Force Survey, for example, indicates that while the foreign-born proportion of total employment in the UK increased from just 7 per cent in 1993–4 to 10 per cent in 2004–5, rates in London for the same period increased from 25 per cent to 34 per cent (Wills et al., 2009). In the least desirable lowest paying jobs, these rates were higher still and as illustrated in Table 16.3, the proportion of foreign-born workers exceeded 60 per cent amongst chefs, cooks, catering assistants and cleaners in 2004–5.

Data like this indicate that cities like London now depend upon foreign-born workers to do the least desirable jobs. Employers are keen to employ migrant workers who take on the labour that locals are often unwilling to do. These findings tally with the arguments made about global cities more generally,

Table 16.3   London's total employment and the proportion of foreign-born labour, by occupation

Occupation	1993–94 (000s)	FB (per cent)	1999–2000 (000s)	FB (per cent)	2001–2 (000s)	FB (per cent)	2004–5 (000s)	FB (per cent)
Chefs, Cooks	29	51	34	61	31	67	38	76
Catering assistants	27	42	25	52	38	55	39	62
Care assistants	22	n/a	41	48	36	38	35	56
Cleaners	64	41	55	46	52	61	51	69
All London	2,894	25	3,262	27	3,349	30	3,375	34
UK	24,449	7	26,687	8	27,114	9	2,7599	10

Note: FB is foreign-born. The reclassification of occupational categories in 2000 means that the data from before and after this date are not directly comparable.
Source: Original analysis of the Labour Force Survey/Annual Population Survey, Wills et al., 2010: 42

whereby Sassen (2001[1991]) has suggested that immigrants will necessarily play a critical part in the labour supply of key city-regions in a globalised world. In brief, Sassen argues that global economic restructuring since the 1970s has led to the emergence of a small number of cities – global cities – that act as important sites of 'command and control' in the new global economy. These cities are key nodes in the world's financial system, and the location of an increasing number of transnational corporate headquarters. Such cities, she argued, experience a rapid growth in the services sector, for both the professions (such as finance, law and associated business services) as well as more routinised jobs in retail, hospitality and personal services. Given the pronounced polarisation in occupational structure and income associated with the services sector, Sassen argued that immigrant labour would be critical in filling the least desirable jobs.

Our research in London has reinforced this analysis, and we have found that surprisingly large numbers of the workers doing 'bottom-end' jobs were born in countries that do not feature in the contemporary immigration regime. In our survey of 429 foreign-born workers employed in low-paid employment in London, we found unexpectedly large numbers of workers from sub-Saharan Africa. Despite the absence of any official immigration channels beyond family reunification and the opportunity for temporary residence as a student, as many as *half* of all the workers we encountered were born in Africa, the majority coming from Ghana and Nigeria (see Wills et al., 2010). African workers had found their way round the British immigration system, arriving as asylum seekers, tourists and students during the past 20 years. Some had since managed to regularise themselves after working illegally, and a small minority of longer-term residents had become British citizens. We also found many migrants from what have been referred to as 'new migrant groups' who have no former ties with the UK through colonialism (Vertovec, 2007). One particularly important 'new migrant group' were Brazilian migrants whom we found working in service sector jobs such as cleaning as well as in construction. Many arrived with student or tourist visas which had expired, leaving them open to exploitation by employers. Initially attracted by the high wages available in London, many Brazilians were keen to return home, especially those from middle-class backgrounds. Given the new immigration regime, future supplies of these types of 'new' and 'old' migrant workers are likely to be greatly reduced. These workers have no presence in the points-based immigration regime and many are now at the sharp end of enforcement in border control.

Issues for future debate

At present, the UK's MDL generates injustice for the migrants who find themselves in low-paid jobs, for those who depend on these workers, and for those who are priced out of employment. Given the changes in the official immigration regime, it is also clear that the government now expects vacancies in so-called unskilled occupations to be filled by new arrivals from the wider EU. This is tending to generate stronger racial divisions between European and non-European workers in the search for low-paid employment within the UK. Whereas our research has found that significant numbers of migrants from poorer parts of the world had arrived to take up employment in a city like London over the past 20 years, often under the radar of the official immigration regime, renewed efforts at border control and workplace surveillance are reducing the opportunities for these types of migrants. Access to the labour market via the asylum system, education, overstaying and unofficial entry is being increasingly restricted by the managed migration regime. This has major consequences for communities in the poorer parts of the developing world, which increasingly depend upon the remittances sent by migrant workers in countries like the UK. Indeed, remittances play a crucial part in determining both why people migrate and where they migrate to. In our own study, we found that the overwhelming majority of migrants remitted money back home, and for those migrants from the poorer parts of the world, this money proved vital in ensuring the survival of their families. Moreover, although the British government now recognises the importance of these remittances and the crucial role they play as 'new finance for development', their simultaneous pursuit of managed migration is endangering these flows by closing down opportunities for migrants from the Global South to migrate to the UK, as well as rendering many of those who are already in the country irregular (Datta et al., 2007). Reinforcing borders around the EU will thus have negative consequences for global development, further entrenching already deep-seated inequalities between the Global North and the Global South.

However, it is also true that migrant labour supplies have consequences for less-skilled workers within the UK – many of whom are settled immigrants or the children of immigrants. The supply of willing foreign-born labour has allowed employers to keep wages low, making these jobs less attractive to 'natives'. In the context of rising unemployment, there is also a growing danger that local workers will blame foreigners for taking 'their jobs'. In a period of economic crisis, it is all too easy to fall into a crude form of protectionism that advocates closing the borders, battening down the hatches, and looking after 'our own'. The next few years will be critical in the evolution of the UK's immigration regime, and we would suggest that attention also needs to be paid to the evolution of the UK's labour market, particularly for the low paid. Improvements in labour standards may yet be critical if rates of unemployment continue to rise and tensions between 'natives' and migrants are to be effectively managed. Improving the quality of low-paid jobs would increase their appeal to 'natives' as well as the migrants currently doing the jobs.

Conclusion

This chapter has outlined the extent to which the social landscape of the UK has been profoundly changed by increased rates – and the increasing diversity – of immigration during the past 20 years. These changes are particularly manifest in the labour market, whereby the increased use of subcontracting and agency labour outlined in the previous chapter has been associated with greater numbers of migrants taking up employment within the UK. In what we are calling new Migrant Divisions of Labour, we can see increased numbers of foreign-born workers at both ends of the occupational hierarchy, and this is particularly evident in a global city like London. In contrast to earlier periods of immigration, many of these workers face restrictions in their access to the labour market and the benefit system as a result of the new immigration regime. Immigration status has itself become an important factor in determining labour market outcomes such that international students, asylum seekers, tourists and irregular workers are to be found doing important jobs like cleaning, care and construction. While the experiences and opportunities faced by these workers often reflect long-standing ethnic and gender divisions of labour, there are also instances in which immigration status can 'trump' discriminations on the basis of race, class and gender. Research in London has found significant numbers of middle-class African men doing jobs that are gendered as female, and eastern European women doing jobs that are generally coded as 'black people's jobs'.

The chapter has illustrated what this means for a city like London, and highlighted the implications it has for communities in the developing world that depend upon remittances, as well as poor communities within the UK. It is clear that immigration both reflects and reinforces patterns of uneven development at all spatial scales. Government policy in the UK at least, looks set to further widen these spatial divides.

Further reading

- Somerville (2007) is an excellent overview of the changes to immigration policy and practice implemented by the New Labour Government in the UK since 1997.
- Wills et al. (2010) provides a more detailed explication of the material presented in this chapter, including additional data and argumentation.
- Winder (2004) offers a populist historical overview of immigration and its effects in Britain.

References

Coe, N.M., Kelly, P.F. and Yeung, H.W.C. (2007) *Economic Geography: A Contemporary Introduction.* Oxford: Blackwell.

Datta, K., McIlwaine, C.J., Wills, J., Evans, Y., Herbert, J. and May, J. (2007) The new development finance or exploiting migrant labour? Remittance sending among low-paid migrant workers in London, *International Development Planning Review,* 29, 1: 43–67.

Home Office (2006) *A Points-Based System: Making Migration Work for Britain,* Paper 6741. London: Home Office.

Home Office and Commonwealth Office (2007) *Managing Global Migration: A Strategy to Build Stronger International Alliances to Manage Migration.* London: Home Office.

Labour Force Survey http://www.statistics.gov.uk/statbase/Source.asp? Link = 358

Legrain, P. (2007) *Immigrants: Your Country Needs Them.* London: Little Brown.

MacKenzie, R. and Forde, C. (2009) The rhetoric of the 'good worker' versus the realities of employers' use and the experience of migrant workers, *Work Employment & Society,* 23, 1: 142–59.

Mathews, G. and Ruhs, M. (2007) *Are you being served? Employer demand for migrant labour in the UK's hospitality sector,* Working Paper. Oxford: COMPAS.

May, J., Wills, J., Datta, K., Evans, Y., Herbert, J. and McIlwaine, C. (2007) Keeping London working: global cities, the British state, and London's new migrant division of labour, *Transactions of the Institute of British Geographers,* 32: 151–67.

Model, S. (2002) Immigrants' social class in three global cities, in M. Cross and R. Moore (eds), *Globalization and the New City: Migrants, Minorities and Urban Transformations in Comparative Perspective.* Basingstoke: Palgrave. pp. 82–118.

Peck, J. (1996) *Work-place.* London: Guilford Press.

Phillimore, J. and Goodson, L. (2006) Problem or opportunity? Asylum seekers, refugees, employment and social exclusion in deprived urban areas, *Urban Studies,* 43: 1715–36.

Piore, M. (1979) *Birds of Passage: Migrant Labor and Industrial Societies.* Cambridge: Cambridge University Press.

Sassen, S. (2001 [1991]) *The Global City: New York, London, Tokyo*. Princeton, NJ: Princeton University Press.

Somerville, W. (2007) *Immigration Under New Labour*. Bristol: Policy Press.

Vertovec, S. (2007) Super diversity and its implications, *Ethnic and Racial Studies*, 30: 1024–54.

Waldinger, M. and Lichter, M. (2003) *How the Other Half Works: Immigration and the Social Organization of Labor*. Berkeley, CA: University of California Press.

Wills, J., May, J., Datta, K., Evans, Y., Herbert, J. and McIlwaine, C.J. (2009) London's migrant division of labour, *European Journal of Urban and Regional Studies*, 16: 257–71.

Wills, J., Datta, K., Evans, Y., Herbert, J., May, J. and McIlwaine, C. (2010) *Global Cities at Work: New Migrant Divisions of Labour*. London: Pluto Press.

Wills, J. with Kakpo, N. and Begum, R. (2009) *The Business Case for the Living Wage: The Story of the Cleaning Service at Queen Mary*. London: Queen Mary, University of London. Available from the author and via www.geog.qmul.ac.uk/staff/willsj.html.

Winder, R. (2004) *Bloody Foreigners: The Story of Immigration to Britain*. London: Little Brown.

Acknowledgements

The authors are grateful to the ESRC for funding and to the editors and publishers of *European Urban and Regional Studies* for permission to reproduce some of the material already published in the journal.

17

THE UK ECONOMY AND THE TRANSFORMATION OF EAST CENTRAL EUROPE

Alison Stenning

AIMS

- To explore the remaking of the UK economy in the context of post-socialist transformations in Europe

- To document the connections between the UK economy and the economies of East Central Europe

- To highlight ways in which the UK's cities and regions have been transformed by 20 years of post-socialism

- To consider how post-socialism transforms our understanding of the geography of the UK economy.

We are all post-socialist now!

On 28 April 2009, just days before the fifth anniversary of Poland's accession to the European Union and some 20 years after the end of communism, Britain's Prime Minister, Gordon Brown, was ridiculed for receiving 'a humiliating lecture on economic competence by the leader of a country that was once one of the poorest in Europe' (*Mail Online*, 2009). The ridicule resulted from the apparently laughable notion that Poland – for it was Poland's Prime Minister, Donald Tusk, who 'lectured' Brown – might be in a position to hand out economic advice to the UK. This moment, and its reception, neatly highlights some questions about the comparative conceptualisation of the British and Polish economies, in economic geography and beyond.

This act of putting Poland's economy in its place suggests a view of the western European economies as dominant and of those in eastern Europe as subordinate and provincial, a view derived from the geopolitics and geo-economics of the Cold War and earlier 'inventions' and 'recycled, celebrated and deployed' (Wolff, 1996: 365) following the collapse of state socialism. From this perspective, whilst it is taken for granted that Poland's economy has been transformed by two decades of UK (and other western) trade and aid, it is barely conceivable that the UK's economic geography might also have been reshaped by its changing relationships with Poland, and the rest of East Central Europe (ECE). Despite some attention paid to the global shifts invoked by the triumph of capitalism and the spread of neoliberalism, economic geographers have overlooked the many ways in which the UK's economic geography has itself been transformed by post-socialism. This omission jars all the more since it has become increasingly commonplace within economic geography to identify the relational spatialities of economic life, that is, to trace the multidimensional networks and articulations of national, regional and local economies, and to explore the ways in which an economy's global connections are created 'from below' as much as from above. In this context, it is increasingly recognised that 'core' economies might be remade by transformations in distant and peripheral economies, yet far more attention has been paid to the UK's transatlantic and post-colonial connections than to its European relationalities, and, in particular, to those which stretch further to the East (Stenning, 2005).

This chapter, then, seeks to highlight the complex and diverse ways in which the economic geography of the UK and of its economic relationships with ECE have been transformed by 20 years of post-socialism. What follows is not primarily a quantitative analysis – though some reference is made to the scale of the processes and effects discussed – but is more focused on the qualitative shifts which suggest a remaking of economic landscapes and relationships. In the following three sections, I review the developing economic connections and take both a chronological and a thematic perspective, as the economic relationships have witnessed a deepening and a broadening over time. I begin by documenting and analysing flows of trade and investment, exploring not only their scale but also their geography. Alongside this, I explore the institutional connections of national and regional co-operation and competition which have paralleled those capital flows. The focus then shifts to migration, and to a discussion of its effects, not only on the labour market but also on institutional change, public services and, importantly, consumption. Bringing together the spheres of business, networks and institutions, I then consider the signs of ECE entrepreneurial and associational life in the UK. In the chapter's concluding section, I turn to an analysis of the geographies of the UK's post-socialism, exploring both the changing spatialities of its external relationships and the uneven development of post-socialism within the UK.

Trade, investment and institutional linkages

Against a background of almost non-existent trade and investment before 1989, UK companies have invested millions of pounds in the new EU member states in the last 20 years and imports and exports have respectively grown to £13.9 billion and £7.7 billion annually (UKTradeInfo, 2009). Whilst UK exports to other EU countries as a whole were characterised by annual growth rates of just 2.6 per cent between 2001 and 2008, exports to Estonia, Slovakia, Poland and Romania grew by 12.6 per cent, 11.0 per cent, 10.6 per cent and 10.0 per cent respectively. Still more marked was the growth in imports from the new member states, especially Slovakia (31.3 per cent), Poland (17.2 per cent), Hungary (16.7 per cent) and the Czech Republic (15.2 per cent), against a background of a general intra-EU import growth of 4.9 per cent. It should, of course, be acknowledged that these growth rates are occurring from the basis of relatively low levels of trade as compared with the UK's major trading partners (such as the US, Germany, the Netherlands and France), but the shifting patterns of trade are important nonetheless.

UK foreign direct investment (FDI) in the new member states includes investments across the full range of economic activity and involves a number of major British companies. In Poland alone, these include BOC, Aviva, Cadbury, Coats Viyella, GSK, HSBC, Provident, Tesco and Tarmac. By the end of 2006, there were 927 firms with British capital in Poland, including 126 whose investments exceeded US$1 million (PAIZ, 2009). Most commonly, these investments are analysed for their transformative effect, positive and negative, on the recipient economies, whilst acknowledging what attracts FDI flows to the East. This downplays the value of such investments to the British firms involved, and to the UK economy as a whole, but also neglects the associated remaking of economic relationships. What may have begun as a search for cheap labour and new markets has developed into much more complex flows of investment, reflecting the complementarities of the British and ECE economies. Indeed, mirroring the changing dynamics of trade, recent years have also seen significant growth in the flows of outward FDI from these economies. Whilst Russian transnationals such as Gazprom and Lukoil are perhaps the most visible examples of eastern investors in the West, other smaller investors from across the region are beginning to appear in the UK and in other western European economies. Some of these flows of East–West investment are connected to migration, but others reflect the growing connections between firms and entrepreneurs in the UK and East Central Europe. In particular, relationships that developed in the early 1990s on the basis of West–East flows of capital and technology have, in recent years, been expanded to form collaborative partnerships and two-way flows of investment (Sawicki, 2009).

Whilst these flows of trade and investment convert to profit for the companies involved, their effects on the UK economy extend beyond that. Some FDI has been driven by a relocation of jobs and production from the UK to Poland and the rest of ECE, with a view to cutting costs. In some cases, threats of relocation appeared more as a rumour than as reality, but some major UK manufacturing plants have been closed or downsized in parallel with expanded investments in the new EU member states. Headline-grabbing examples include Electrolux, which closed its Spennymoor, County Durham, plant in late 2008 with the loss of 500 jobs and moved production to Świdnica, Poland (McKay, 2007), and Black and Decker, which cut jobs and production at plants in County Durham and North Yorkshire whilst simultaneously expanding investments in the Czech Republic (*Northern Echo*, 2008). The primary motivation for these relocations was seen to be an increasingly competitive market which was eroding the viability of relatively high-cost locations in the UK, and local trade unionists and politicians highlighted their concerns that UK jobs were being replaced by jobs in cheaper, more eastern locations. Whilst such relocations were evident throughout the 1990s and early 2000s, a renewed wave seemed to occur a few years after accession, perhaps as a result of accession itself, but also, it seems, provoked by the positive experience manufacturers have had in employing Central European migrants in their UK plants.

In these ways, communities in the UK, often those on the margins of the UK economy which had already been hit hard by earlier rounds of disinvestment and deindustrialisation, have found themselves in competition with their new European neighbours for jobs and investment. Multinationals' search for lower-cost locations was eased by the accession to the EU of eight Central European states with apparently attractive investment climates: skilled but low-paid workforces, increasingly good access to both domestic and wider European markets, and common quality standards and legislative frameworks. Job loss, shrinking labour markets and declining investment in Spennymoor, for example, can be seen to be a direct effect of the transition to capitalism in ECE.

This shift to the East can be seen in other spheres too, as the UK's Northeast acknowledged in its submission to the debate over the future of the EU's structural funds. Not only have policy actors in the region raised concerns about the knock-on impact of the costs of pre- and post-accession aid in the East and about the changing position of Objective 1 regions in the 'old' EU as the 75 per cent funding threshold inevitably fell lower, but they also noted that, with accession, 'the centre of gravity of the EU will move East' (European Management Board for the North-East, 2002). As the EU enlarges to the East and South-east, the UK's peripheries become still more peripheral. For some cities and regions, the response to both these processes of change – competition for investment and the financial and territorial shift to the East – has been to build collaborative relationships with their eastern neighbours, in part to share their experience of restructuring and regeneration, but also to promote shared interests and to counter, if only to a limited extent, the enforced competition for investment. Much of the early contact between eastern and western regions was shaped by EU and UK schemes for the transfer of know-how from West to East and reflected an uneven geography, but more recently relationships have become more equal. The varied forms such relationships take echo earlier experiences of twinning and international co-operation but are also employed to imagine new European geographies, which seek to counter the peripherality of many regions, East and West. For example, Newcastle City Council's international partnerships with Malmo,

Tallinn and Gdańsk enable not only learning across cities with common economic histories but also connections within a northern, Baltic Europe which circumvents Europe's core.

Migration, labour markets and consumption

The focus on trade, investment and institutional linkages has been overtaken somewhat since 2004 by the migrations which followed the accession of eight ECE states to the EU (and two more in 2007). It is important to note that post-2004 migration was just the most recent in a series of flows from ECE to the UK. In addition to late-nineteenth and early-twentieth century migrations, the post-war settlement of refugees and displaced persons resulted in over 400,000 East European workers settling in the UK between 1947 and 1951, most often in regions with unmet demand for labour. Throughout the post-war period, intermittent waves of migrants arrived from the East, such that by 2001 there were approximately 180,000 ECE-born people resident in the UK. In addition to these permanent residents, seasonal workers from ECE were also playing a major role in the UK economy. By the early 2000s, the overwhelming majority of the 100,000-plus workers coming to the UK annually through the seasonal agricultural workers scheme were from the new member states (Clarke and Salt, 2003), and many more were working in the UK on self-employment visas (Garapich, 2008).

Nevertheless, the opening of the UK's labour market to citizens of the new member states in May 2004 did mark a step change in the scale of migration flows. By the end of 2008, some 926,000 applications to the Worker Registration Scheme – established to monitor post-accession migration – had been approved (Border and Immigration Agency, 2009a). In addition, in 2007 and 2008 some 80,000 Romanians and Bulgarians were authorised, through various schemes, to work in the UK (Border and Immigration Agency, 2009b). In total, between 2002 and the end of September 2008, there had been 1,242,200 applications for National Insurance Numbers (NINOs) from citizens of the ten new member states, applications which included not only employees but also the self-employed, including the archetypal Polish plumber, contracted workers in construction and security, and growing numbers of ECE entrepreneurs.

The debate about the impact of these migrations on the UK economy has focused on macroeconomic assessments of gross domestic product (GDP), inflation and productivity (for a review, see Stenning and Dawley, 2009). Since this is such a key political issue, it is unlikely that there will ever be a definitive assessment, but it is clear that in certain sectors of the UK economy, such as construction and agriculture, ECE migrant workers are making a significant contribution. This contribution rests not only on the increased availability of workers to plug hard-to-fill vacancies, but also on the willingness of migrants to take on work with poor conditions and low rates of pay, and concerns have been raised not only for the migrant workers themselves, but also for indigenous workers seeking entry-level employment. Alongside debates about the macroeconomic impact of migration, then, attention has also been drawn to the impact of immigration on wages, unemployment and workforce development within the 'native' workforce. Whilst there is no consistent evidence of such an impact, there is some evidence that post-accession migration has affected low-paid labour markets, 'displacing' young workers and those in localities afflicted by worklessness, many of which have already lost manufacturing jobs to the new member states. Moreover, in addition to meeting relatively unskilled labour shortages, there are dentists, doctors and nurses who have been actively recruited from ECE, reinforcing the reliance of the UK public sector on migrant employment. In more intangible ways, migrant workers can be seen to be 'importing' working practices from ECE to UK workplaces and 'exporting' their new skills and experiences home, remaking work in both places.

In the UK nations and regions, considerable policy attention has been paid to these new migrant flows and to 'capturing' their economic potential. These responses reflect both a proactive and strategic interest in tackling the challenges of productivity, skills and population decline and a more reactive response to the unexpected influx of migrants. Many of these initiatives directly draw inspiration from the Scottish Executive's Fresh Talent Campaign, which has focused on in-migration as the key to Scotland's future demographic and economic vibrancy, and 'many areas have undertaken studies to assess the economic benefit of migration and stress that migrants are over-represented in hard-to-fill vacancies and are helping to fuel local economic growth' (Local Government Association, 2007a: 5). It is not only local authorities who have reacted to this new migration, but also labour market intermediaries, from trade unions to employment agencies. There has been a noticeable expansion of new, specialist agencies, often working in partnership with agencies and recruiters in ECE, and multinational agencies have also increasingly turned their attention to this new supply of labour (see Figure 17.1). In the Newcastle branch of one national agency, for example, managers have built on their success in recruiting and placing post-accession migrants to expand their business, establish new client contracts, and increase the numbers of workers they place in client workplaces, demonstrating clearly the importance of this new labour force for this sector, and for its economic geographies (Stenning and Dawley, 2009). Within trade unions, the immediate post-accession months were focused on a concern that migrant workers offered employers a non-unionised and compliant workforce which could be used to disadvantage organised, indigenous workers, but the focus of attention soon shifted to organising migrant workers. Not only have migrant workers been brought into mainstream branches, but new forms of organising have been introduced for migrants, which seek to reflect migrants' transnational working lives, through partnerships with Polish unions, for example. These new forms of

Figure 17.1 An employment agency seeks to attract Polish workers
(Photo © A. Stenning)

organising, borrowing from previous attempts to organise non-traditional workforces and to organise in the scales and spaces of workers' lives, suggest an important shift in the geography of British trade unions and their international relationships (Fitzgerald and Hardy, forthcoming).

Increasingly, however, ECE migrants are being seen not only as workers but as consumers too, as the value of the 'Polish pound' (or, more accurately, the East European pound) has been recognised. In 2008, Mintel estimated that the annual spending power of post-accession migrant workers in the UK had risen to some £8.4 billion (*Financial Times*, 2008) and was therefore of the same scale as East–West trade. Despite the fact that many are employed in low-paid jobs, their dominant demographic and lifestyle characteristics, and the fact that fewer than expected are remitting income home, suggest that their disposable income is disproportionately high; when in 2006 the 'Polish pound' was valued at just £4 billion, this was equated with 'adding the consumer demand of Liverpool to the economy' (*The Telegraph*, 2006). It is not surprising that more and more British firms are seeking to target this valuable market. Tesco, Asda and Sainsbury's all now stock Polish food in the majority of their UK stores, and for Tesco it is the 'fastest-growing ethnic food range ever launched in Britain' (*Mail Online*, 2007). Heinz and Nestlé have both started distributing their Polish-brand products in the UK; sales of Polish beer have significantly boosted the economic health of both brewer SABMiller and pub chain JD Wetherspoon; WHSmith and Borders now stock Polish-language books; DIY firms B&Q and Wickes are producing advertisements and flyers in Polish; British high street banks have launched specialised Polish products and employ Polish staff; and a vast range of other services are being developed to serve these new markets. Importantly, not only do these new consumers offer a valuable market for British firms, especially valuable in times of economic crisis, but these products are also increasingly being explored by British consumers – Tyskie, bigos and vodka have the potential to change British patterns of consumption as Indian and Chinese cuisine did in previous generations. In some cases – such as Lloyds TSB's money transfer card – product innovations originally developed for the migrant worker market have been taken up more widely by other transnational users, such as gap-year students.

In addition, however, to the appearance of new markets and new products, these attempts to serve migrant consumers suggest other economic-geographical shifts too. Tesco, Heinz, Nestlé and SABMiller have been particularly successful in reaching these consumers because they were already very active in Polish markets. Tesco, for example, possesses established Polish supply chains which already serve its Polish stores, and this has enabled them to move swiftly and easily to export to the UK. Those newer to the Polish market, such as Borders and NatWest, have sought relationships with Polish partners to identify key products and to develop targeted marketing strategies. In these ways, the development of migrant markets supports new economic relationships within and between firms and suggests more complex relationships than those established in the 1990s. It is also important to note that the response of both UK and global firms to these new migrant markets has been much swifter and more pronounced than responses to earlier waves of migration. In part this might be explained simply by the scale of migration, but there are perhaps other explanations; the timing of their migration has paralleled the growing integration of European corporate economies, suggesting that some UK firms have been particularly well placed to serve these markets, and to use them as an entry point for further activity in ECE.

Post-accession migrants are not only consuming commercial goods and services, however, but also public services, such as housing, schools, the health service and public transport, provoking fears of 'overcrowding' and budget pressure at local and national scales. For some local authorities, these concerns were exacerbated by the apparent spatial mismatch between the costs and benefits of migration, as government funding for essential services such as interpreters, support workers, and bilingual teaching assistants was allocated on the basis of outdated and inadequate population data. In some towns and cities, moreover, post-accession migrants have settled predominantly in areas of existing disadvantage, stirring even greater anxieties amongst service providers. In raising these concerns, the Local Government Association (2007b) highlighted a problematic economic geography of migration, acknowledging that migration was having a positive impact on the UK economy but arguing that 'the money that is being generated isn't necessarily finding its way back down to the local level'.

Figure 17.2 An entrepreneur seeks to meet Polish and Czech demands (Photo © A. Stenning)

Enterprise and business networks

The growth of migrant 'wants and demands' provokes a response not only from mainstream providers, but also from 'ethnic' enterprises, established within migrant communities to serve particular interests. The growth of a post-accession 'migration industry' (Garapich, 2008) incorporates media outlets, language and skills trainers, recruitment agencies, transport and money transfer services, health centres, benefits and financial advisors, wholesalers and retailers, and producers. Some of these emerged out of the informal and individual activities of earlier migrants; others have developed as a 'critical mass' of co-ethnics has settled in the UK's cities and regions (see Figure 17.2). Garapich (2008) notes that this developing industry extends beyond an 'ethnic niche' to incorporate many of the mainstream providers highlighted above, as well as serving neighbouring ethnic groups. Established Polish migrant businesses, in particular, are increasingly targeting less well-served communities, such as Czechs and Slovaks, and seeking to capitalise on the most recent inflows of Bulgarian and Romanian migrants, many of whom lack the established networks of the British–Polish community.

There is little doubt that for many the migration industry is lucrative, but it is not the only industry in which migrant entrepreneurs are active. As I have already suggested, self-employed workers make up a considerable share of those who have arrived since (and before) 2004, establishing businesses in sectors such as construction, security and cleaning. Whilst some of these businesses reflect a problematic 'contracting-out' of employment in established workplaces as employers reject the responsibilities and costs of employment, others suggest that settled migrants are establishing small enterprises employing co-ethnics, and/or other East Central Europeans. In contrast to those discussed above, these firms rest not on co-ethnic markets but on demand in the wider economy for the skills and 'reputation' of ECE workers.

These signs of entrepreneurial activity have attracted the attention of commercial, statutory and voluntary business support organisations. Whilst high street banks such as HSBC have launched specialist business support services for Poles, organisations such as Business Link, the regional development agencies, and the British Polish Chamber of Commerce, are developing workshops, courses and information for would-be entrepreneurs, reflecting a recognition that migrant enterprise can play an important role in the economic life of cities and regions, especially within marginalised neighbourhoods and localities.

To this end, it is also important to draw attention to the associational life of central European migrants, in itself integrated into the migration industry described above, but also increasingly reflecting the agency of Polish and other migrants in British public life. Most UK cities, and many small towns too, now have active Polish community organisations which act not only as sites of support and socialising but also as the focal point of ever-expanding activities, such as the production of newsletters and the development of Polish-language schools. Importantly, this associational life also includes networks oriented explicitly towards to the support of economic activity, such as Polish Professionals in London, the British Polish Business Club, and the Association of Polish Entrepreneurs and Companies UK. Such organisations, though often small-scale and localised, are new economic actors involved in transnational activities, such as sharing skills and know-how, supporting the establishment of Polish small- and medium-sized enterprises in the UK, facilitating the process of investing back home, and advising British firms on accessing Central European markets, in the UK or in the East.

The value of such key actors, who work between business environments in the East and West, has not been lost on government departments and employers' organisations. In the early 2000s, UK Trade and Investment drew attention to the role that émigré Central Europeans might play for the UK as 'export promoters' by offering language skills and economic and cultural knowledge of these new trading partners and, more recently, employers have highlighted the potential contribution that well-educated, technically-trained employees with foreign-language skills might make to the expansion of British business in the new member states.

Conclusions: the economic geographies of post-socialism in the UK

In all of these ways, the post-socialist transformation of ECE can be seen to be having an impact on the UK economy. These emerging features of British economic life point to particular economic geographies, which transform the economic landscapes of UK cities and regions and which call attention to new economic relationalities. Still other examples – of low-cost airlines, new media, UK pension investments in Central European property portfolios, or trafficking and organised crime – could extend the argument further.

UK cities and regions find themselves competing with cities and regions in ECE for investment, for labour and for structural funds, for example, but also build new forms of co-operation with partners previously seen as distant and disconnected, and in this way envisage new European geographies. The communities and neighbourhoods of those cities and regions are now home to new minorities – there are post-accession migrants living in every single local authority in the UK – who are working alongside British workers, who are bringing with them new products and services, and who are, in some cases, creating transnational households whose economies stretch across borders. For UK firms, government agencies, trade unions and other institutions, the impacts of post-socialism have increasingly been seen in the formation of new economic geographies which extend their activities and relationships to new places and new partners, and which complicate the flows from West to East which were seen to dominate in the early 1990s – and which are often still assumed to be the most significant manifestations of post-socialism in the West. And new economic actors have emerged, such as migrant entrepreneurs, associations and networks, who seek to cement economic relationships between the UK and their ECE homes and whose economic activities rest on the transnationality of their lives.

The multidimensionality of these flows is increasingly evident, as goods, services, capital and people move according to varied temporalities and geographies, and point to a future in which movements between East and West, facilitated by common citizenship rights, common currencies and low-cost travel networks, are a common feature of Europe's economic geographies (Stenning and Dawley, 2009). Whilst it is perhaps too early to tell how patterns of migration will settle in the future – and recent reports suggest that flows are reversing – it is possible to highlight some of the benefits to the UK economy of post-accession mobility. Not only are these evident in the increasing familiarity with (and attachment to) British firms, products and services, but also in the presence of a cohort of Poles, Czechs, Slovaks and so on who have experience and an understanding of British culture, British business practice and the English language, and who have relationships which might support the expansion of transnational enterprise, trade and investment.

Another important aspect of the economic geographies of post-socialism in the UK is that they touch most parts of the country, from London, through provincial cities such as Newcastle and Birmingham, to small towns and villages in Cornwall and the Highlands. But this extensive presence belies an uneven development. Whilst firms headquartered and located across the UK have engaged in trade and investment in ECE, it is those cities and regions on the periphery which have been most disadvantaged by disinvestment. The impacts of migration have been uneven too, such that greater numbers of migrants have arrived in London and the South-east, contributing particularly to London's division of labour (see Chapter 16), feeding a concomitant clustering of migrant enterprise, and, perhaps, exacerbating the region's housing and public service pressures. On the other hand, the presence of migrants across the UK, often in places which had seen little earlier in-migration, has created significant challenges for local authorities and service providers without a track record in managing migration. The ubiquitous geographies of post-accession migration are very different to previous waves of migration to the UK and this, in itself, is an important economic, and social, geographical phenomenon.

It is difficult to predict how these processes will play out over the next 5 or 20 years, but it is possible to claim that the passage of post-socialism – and of EU enlargement – has created new avenues for development and new spaces of economic life in the UK. This suggests that there is a need to reflect more fully on the post-socialist economic geographies of the UK and to acknowledge that the UK economy has been remade by the apparently provincial transformations of post-socialism, a need which becomes all the more urgent in the context of the ongoing economic crisis. The crisis challenges the policies and practices of neo-liberalism, which emerged alongside – and in part from – the post-socialist transformation of Europe, and the uneven development of the crisis, as the Tusk–Brown encounter described above suggests, remakes the economic geographies of Europe once more. This, in turn, makes economic geography's provincialising of post-socialism increasingly untenable and challenges economic geographers to explore these new economic relationships in much more detail.

Further reading

- Burrell (2009) offers an excellent introduction to recent research on Polish migration to the UK.

- Henry et al. (2002) explore the idea of 'globalisation from below' and how migrant and minority ethnic communities can play a role in reshaping economic geographies.
- Stenning and Hörschelmann (2008) debate the meanings and geographies of post-socialism.

References

Border and Immigration Agency (BIA) (2009a) *Accession Monitoring Report, May 2004– December 2008*. London: BIA.

Border and Immigration Agency (BIA) (2009b) *Bulgarian and Romanian Accession Statistics, October–December 2008*. London: BIA.

Burrell, K. (ed.) (2009) *Polish Migration to the UK in the 'New' European Union*. Farnham: Ashgate.

Clarke, J. and Salt, J. (2003) Work permits and foreign labour in the UK: a statistical review, *Labour Market Trends*, November: 563–74.

European Management Board for the North-East (2002) *North-East Regional Position on the Future of EU Cohesion Policy*. Available at http://ec.europa.eu/regional_policy/ debate/document/futur/member/north_east_of_england_sep02.pdf, accessed 6.5.09.

Financial Times, The (2008) Battle intensifies to capture 'Polish pound', 9 September. Available at www.ft.com/cms/s/0/5d55d532-7df7-11dd-bdbd-000077b07658.html, accessed 6.5.09.

Fitzgerald, I. and Hardy, J. (forthcoming) Thinking outside the box? Trade union organising strategies and Polish migrant workers in the UK, *British Journal of Industrial Relations*.

Garapich, M. (2008) The migration industry and civil society: Polish immigrants in the United Kingdom before and after EU enlargement, *Journal of Ethnic and Migration Studies*, 34: 735–52.

Henry, N., McEwan, C. and Pollard, J. (2002) Globalization from below: Birmingham – postcolonial workshop of the world?, *Area*, 34: 117–27.

Local Government Association (LGA) (2007a) *Estimating the Scale and Impacts of Migration at the Local Level*, Available at www.lga.gov.uk/lga/aio/109536, accessed 6.5.09.

Local Government Association (LGA) (2007b) Contingency fund needed to help councils deal with migration, *LGA News Release*, 1 November. Available at www.lga.gov. uk/lga/core/page.do?pageId=41570, accessed 6.5.09.

Mail Online (2007) Polish food sales soar ten-fold, 16 April. Available at www.daily-mail.co.uk/news/article-448878/Polish-food-sales-soar-fold.html, accessed 6.5.09.

Mail Online (2009) Stony-faced Brown gets ANOTHER lecture on the economy ... from Poland's prime minister, 28 April. Available at www.dailymail.co.uk/news/world-news/article-1174380/Stony-faced-Brown-gets-ANOTHER-lecture-economy-Polands-prime-minister.html, accessed 6.5.09.

McKay, N. (2007) 500 jobs to go as Electrolux closes, *The Journal* (Newcastle), Available at www.journallive.co.uk/north-east-news/todays-news/2007/12/15/500-jobs-to-go-as-electrolux-closes-61634-20252547/, accessed 6.5.09.

Northern Echo (2008) Region suffers as bosses move manufacturing plants to Eastern Europe, 30 January. Available at http://archive.thenorthernecho.co.uk/2008/1/30/241354.html, accessed 22.6.09.

PAIZ (2009) *Foreign Direct Investments in Poland.* Available at www.paiz.gov.pl/index/?id=59112692262234e3fad47fa8eabf03a4, accessed 6.5.09.

Sawicki, S. (2009) *British–Polish Networks and the New Economic Geographies of Europe,* PhD Thesis, Newcastle University.

Stenning, A. (2005) Out there and in here: studying Eastern Europe in the West, *Area,* 37: 378–83.

Stenning, A. and Dawley, S. (2009) Poles to Newcastle: grounding new migrant flows in peripheral regions, *European Urban and Regional Studies,* 16: 273–94.

Stenning, A. and Hörschelmann, K. (2008) History, geography and difference in the post-socialist world; or, do we still need post-socialism?, *Antipode,* 40: 312–35.

Telegraph, The (2006) UK plc targets the Polish pound, 10 September. Available at www.telegraph.co.uk/finance/2947064/UK-plc-targets-the-Polish-pound.html, accessed 6.5.09.

UKTradeInfo (2009) *Summary of Trade with EU Countries 2001 to 2008.* Available at www.uktradeinfo.com/pagecontent/datapages/tables/EU_NonEU_Ann08.XLS, accessed 6.5.09.

Wolff, L. (1996) *Inventing Eastern Europe.* Stanford: Stanford University Press.

18

CODA: THE UK ECONOMY IN AN ERA OF GLOBALISATION

Neil M. Coe and Andrew Jones

AIMS

- To draw together the major themes that emerge from the different aspects of the UK's economic development covered in this book

- To consider how key processes affecting the UK economy continue to produce uneven economic development

- To highlight the key challenges faced by policy makers in the next decade

The UK economy: an ongoing story of persistent uneven development

At the end of the first decade of the twenty-first century, the UK economy appears to have reached something of a crossroads. The financial crisis of the late noughties precipitated a recession that – as the International Monetory Fund (IMF) commented at the time – undoubtedly more seriously affected the UK economy than other advanced industrial economies. In Europe, for example, Germany, France and even Italy fared better than the UK, entering shallower recessions and escaping sooner. By contrast the UK economy, it seems, has emerged from the global economic downturn as a laggard, and its capacity to bounce back to the levels of either growth or output seen in the noughties boom is by no means ensured. Clearly, a significant contributor to this situation has been the dominance of banking and finance in the UK's economic base, and the future of these industries has been the subject of much debate in recent times. However, in bringing this book to a conclusion, we want to argue that the various contributions to this collection reveal a more serious and underlying story of persistent uneven economic development and that the

UK's recent economic performance cannot just be explained by rather narrow reference to its exposure to an unusually serious financial crisis.

To do this, we want to highlight four innately geographic themes that recur through many of the chapters in this book. The first of these is *regional inequality*, which clearly remains a central feature of the UK's economic landscape. The discussion in Part I provides a detailed examination of how social and economic inequalities continue to underpin major differences between the regions of the UK. The North–South divide remains a persistent feature of UK economic development, with London and the South–east continuing to dominate in gross domestic product (GDP) and growth terms much as they did in the 1980s. Likewise, in Part II analysis of trends in the UK housing market only serve to reinforce this point, as does the evidence concerning the centrality of business services in the UK economy with their London and the South-east focus, or conversely, Part III's examination of the ongoing decline of Britain's manufacturing industry which has long characterised the economic base of other regions. Despite 20 years or more of regional economic development policy initiatives, urban regeneration schemes and attempts to attract foreign direct investment, the processes of *globalization, deindustrialisation* and *tertiarisation* that we identified in Chapter 1 have, since the late 1980s, continued to shape the UK's economic landscape in a distinctively uneven way, concentrating economic activity and power in the South-east.

This leads to a second and closely interrelated theme that emerges from many of the chapters: the *imbalanced nature of the UK's economic base*. The argument about the dangers of uneven regional economic development made in Part I are developed in different ways by other contributions to the book. In Part II, several chapters outline the problems the UK economy faces in terms of its heavy reliance on financial industries, both in terms of the power of the City of London and the vulnerability of provincial cities with respect to the composition of their financial service sectors and their need to compete with each other. Part III also provides analysis of how *deindustrialisation* has continued to weaken the capacity of the UK economy to manufacture both for domestic consumption and export, a feature that other European economies have to a greater degree managed to avoid. Part III also reveals both complexity and further imbalances in the contemporary makeup of other key industries in the UK economy, notably agriculture and retailing.

A third theme is *labour market transformation*. In the introductory chapter we pointed to the fact that over the last 20 years processes of *flexibilisation* and *immigration* have dramatically altered the nature of labour markets and the experience of work in the UK. Part I again provides the context for understanding the broad pattern of labour market inequalities that continue to exist in the contemporary UK economy, but it is in Parts III and IV in particular that the detail and nature of this transformation and its impact on different regions and the people who work in them is revealed. Here the dramatic transformation of the kinds of jobs that the UK economy now creates compared to two decades ago also leads to the argument that *neoliberalization* has heightened labour market inequalities which also reinforce social inequality. Furthermore, as the discussion of the impacts of European scale immigration reveals, the UK's economic development is increasingly tied into a new and complex Migrant Division of Labour (MDL) with the development of key industries and regions such as London becoming increasingly reliant on migrant workers who fill low-paid, 'undesirable' jobs. Part IV also shows how the UK economy's fate is increasingly entwined with the wider European Union.

Finally, a fourth theme that runs through the chapters in the book concerns the nature of *governance and regulation*, along with the way these produce uneven economic development. In Part II, the role of inadequate regulation in banking and financial services is shown to be important in understanding the development of the 2007–9 financial crisis. Yet moving beyond London and the South-east, Part II also examined how UK government expenditure plays a crucial role in shaping economic development in much of the UK's economic landscape. The government's active role in shaping economic geographies was also discussed in relation to the tensions created by regional political devolution in the UK over the last decade, with the argument being that regional devolution has dissolved into a wide range of 'bottom-up' local economic state initiatives aimed at promoting development – a process that has seemingly not had the effect of reducing geographical inequalities.

The next 20 years: challenging times for the UK?

We want to end this book by briefly considering what we suggest to be some of the major challenges facing businesses, policy makers, workers and other stakeholders in the UK economy. We argue that it is important in considering how to face these challenges that policy is conceptualised from a geographical perspective, since too much of the discussion about UK economic policy remains blind to the uneven implications of various initiatives. We want to suggest that there are at least four major challenges that face policy-makers in relation to the UK economy over the coming decade. In many cases these relate to managing the tensions and processes that we have just identified.

First, as the analysis in this book clearly demonstrates, the UK economy remains highly uneven in territorial terms and this continues to produce social inequality and environmental problems for the UK as a whole. Despite some efforts by policy makers to redress the regional imbalance of economic activity in the UK, the North–South divide remains an enduring feature of the UK economic landscape. On the one hand this therefore continues to present overcrowding and shortage problems (of land, labour, skill etc.) in London and the South-east, and on the other the prospects for many northern regions of the UK continue to be at best mixed. Urban deprivation, unemployment and low-paid, low-skill jobs remain key policy imperatives in many parts of the UK in the post-recessionary context, and it seems that past policy approaches – however successful or unsuccessful they were deemed to be on the surface – have failed to tackle the root causes of these problems. The continuation of policies that are either blind to these differentials, or indeed actively seek to reinforce inequalities (e.g. Leunig and Swaffield, 2008), are likely to be increasingly socially and politically divisive.

Second, in terms of the industrial base of the UK economy, the ongoing viability of an economy so heavily dependent on financial and business services has to be questioned. That is not to say that London's role as a global city should not be a priority, because clearly its highly strategic position in the global economy is a major strength of the UK. However, whilst deindustrialisation and manufacturing has affected other major economies including the largest EU economies (France, Germany, Italy), these economies continue to out-perform the UK and weather the storm of globalisation more effectively. It is highly questionable whether the UK can retain its position as one of the world's largest economies without competing more effectively in manufacturing, and other countries have arguably implemented more effective policies than the UK over the last 20 years in terms of preserving the health of their manufacturing sector.

A third policy challenge is the recent global shift towards new environmental industries, sustainable energy and a low-carbon economy. As the chapters in Part III of this book discuss, the UK has begun to move in this direction but in many ways has again lagged behind other large economies. If the UK's energy needs are to be met in a sustainable manner, and if it is to compete in these key industries in the global economy in the coming decades, then there needs to be a significant set of policy incentives and support structures to facilitate this in the UK economy. While discussions of a Green New Deal, for example, show promise (New Economics Foundation, 2008), truly joined-up and progressive policy intervention in these areas still appears distant and is another domain in which UK policy seems to lag behind other major European economies.

Fourth, policymakers continue to face significant labour market challenges. The politics of immigration over the last decade have exposed a contradiction in the way in which the economy's labour needs are met, as well as the quality of the jobs being created. If the UK economy is to remain globally competitive, then the politics of immigration may act as a constraint on the capacity of industries to recruit either sufficient numbers or sufficiently skilled workers. Equally, in terms of social inequality, *tertiarisation* and *neoliberalisation* have produced a UK economy that now employs a growing proportion of workers in casual, temporary and insecure jobs, leading both to lower work–life satisfaction and a longer-term lack of suitable skilled and motivated employees. Other major European economies arguably have managed the impacts of globalisation and neoliberalisation on their workforces more effectively, and this therefore remains an issue for UK policymakers to engage with.

We conclude by reprising two geographical arguments with which we opened the book. First, the chapters in this volume have clearly demonstrated the added-value of approaching the UK economy from a geographical perspective. Approaches which treat the UK as a uniform space run the risk of missing the significant economic, social and political inequalities that shape our everyday lives. Second, the book has illustrated how, now more than ever before, understanding the evolving uneven geographies *within* the UK economy can only be understood through analysis of the UK's shifting position in wider European and global divisions of labour. Researchers and policymakers alike need to grapple with this realisation if they are to intervene effectively in redressing the profound geographical and social imbalances that continue to characterise the UK economy.

References

Leunig, T. and Swaffield, J. (2008) *Cities Unlimited: Making Urban Regeneration Work.* London: Policy Exchange.

New Economics Foundation (2008) *A Green New Deal: Joined-up Policies to Solve the Triple Crunch of the Credit Crisis, Climate Change and High Oil Prices.* London: NEF.

INDEX